Phoenix Iron Company

Useful Information for Architects, Engineers and Workers in

Wrought Iron

Phoenix Iron Company

Useful Information for Architects, Engineers and Workers in Wrought Iron

ISBN/EAN: 9783337395537

Printed in Europe, USA, Canada, Australia, Japan

Cover: Foto ©berggeist007 / pixelio.de

More available books at **www.hansebooks.com**

USEFUL INFORMATION

FOR

ARCHITECTS, ENGINEERS,

AND

WORKERS IN WROUGHT IRON,

BY THE

PHŒNIX IRON COMPANY.

OFFICE,

410 WALNUT STREET, PHILADELPHIA.

WORKS,

PHŒNIXVILLE, PA.

REVISED EDITION, 1886.

PRINTED BY
J. B. LIPPINCOTT COMPANY,
PHILADELPHIA.

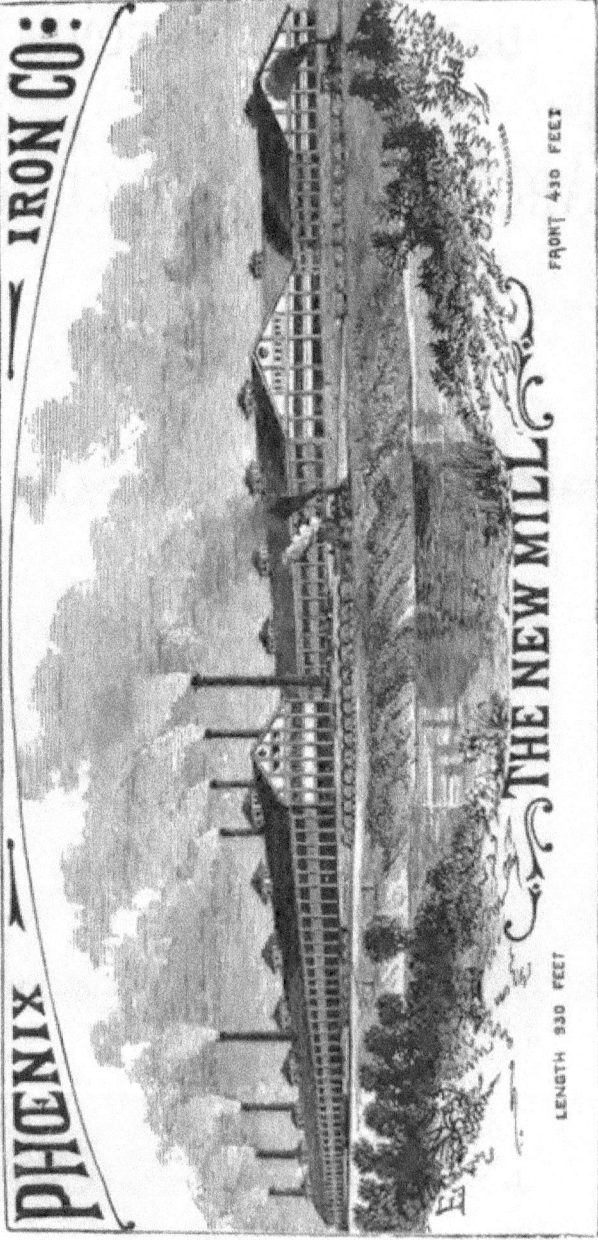

PHŒNIX ← IRON CO: →

THE NEW MILL.

LENGTH 930 FEET

FRONT 430 FEET

INTERIOR OF MACHINE SHOP.

⋙ OFFICERS. ⋘

DAVID REEVES, President,

GEORGE GERRY WHITE, Secretary,

JAMES O. PEASE, Treasurer,

PHILADELPHIA.

W. H. REEVES, General Superintendent,

AMORY COFFIN, Chief Engineer,

R. H. DAVIES, Master Mechanic,

PHŒNIXVILLE.

Correspondents will please address

PHŒNIX IRON COMPANY,

410 Walnut Street,

PHILADELPHIA.

BLAST FURNACES.

CONTENTS.

SHAPES of WROUGHT IRON

Manufactured by the

PHŒNIX IRON CO.

PHILADELPHIA,

PA.

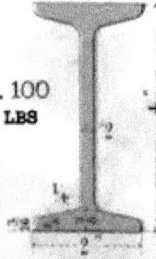

No. 100
18 LBS

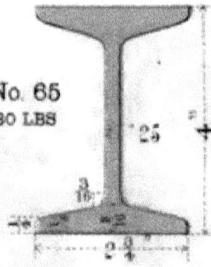

No. 65
30 LBS

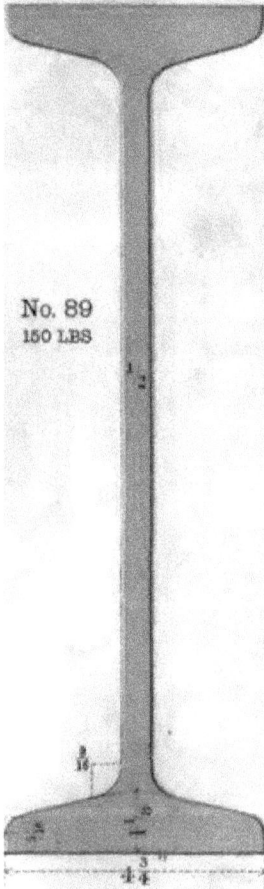

No. 89
150 LBS

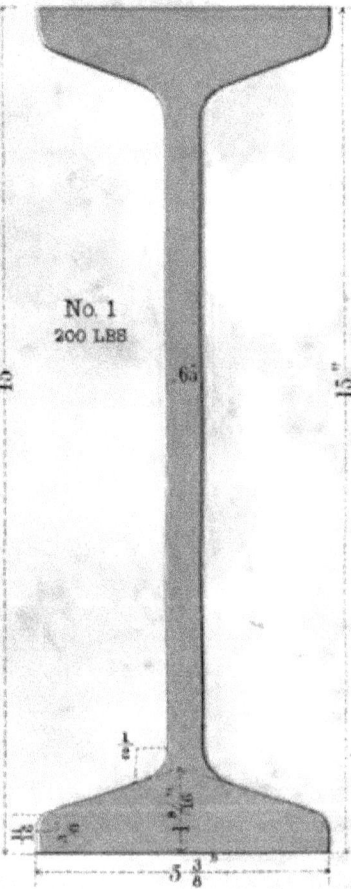

No. 1
200 LBS

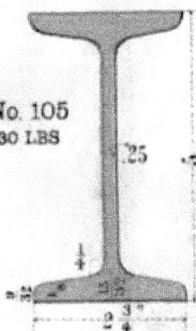

No. 105
30 LBS

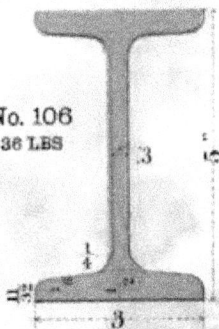

No. 106
36 LBS

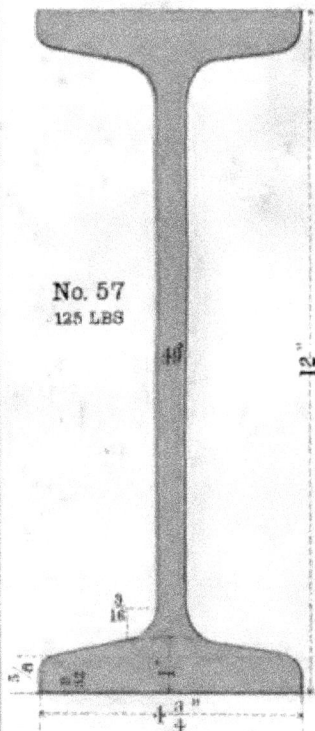

No. 57
125 LBS

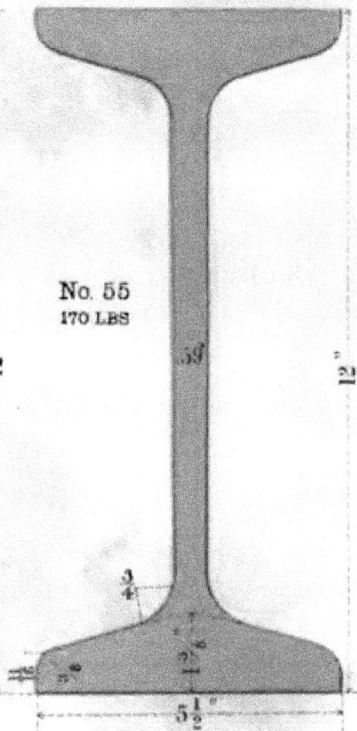

No. 55
170 LBS

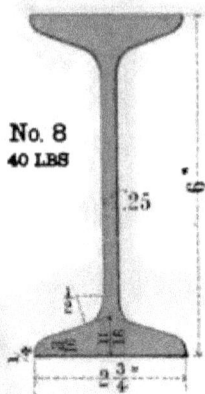

No. 8
40 LBS

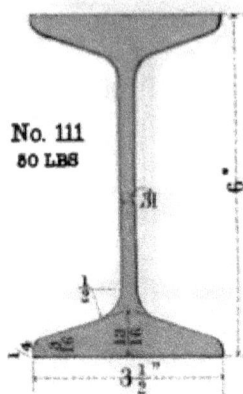

No. 111
50 LBS

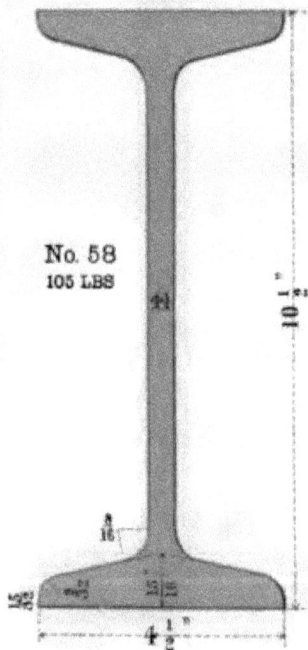

No. 58
105 LBS

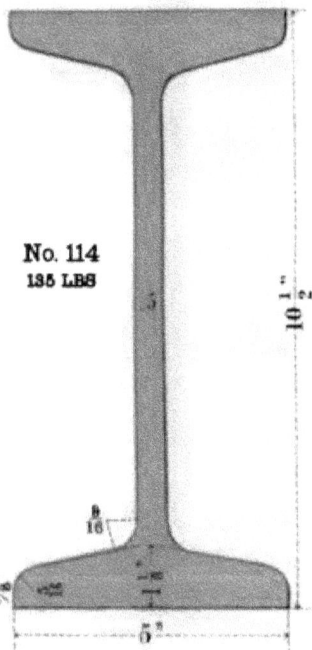

No. 114
135 LBS

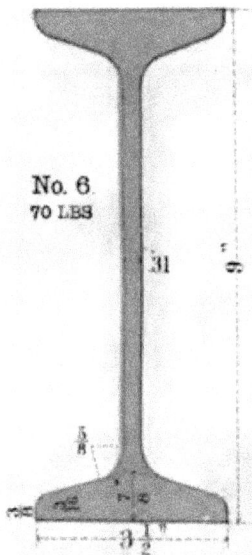

No. 6.
70 LBS

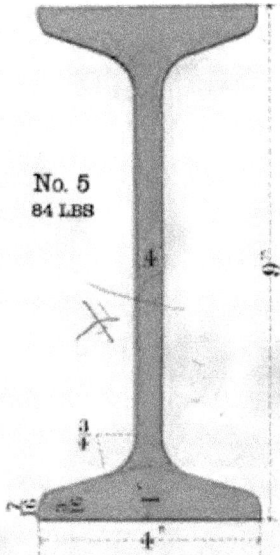

No. 5
84 LBS

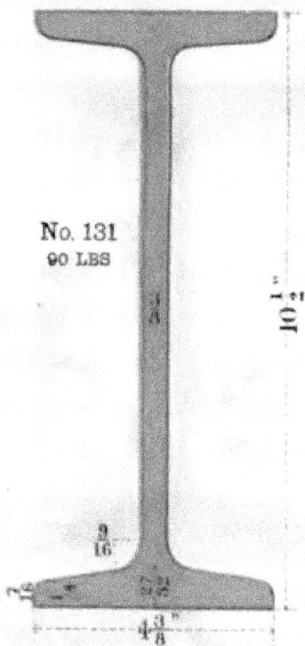

No. 131
90 LBS

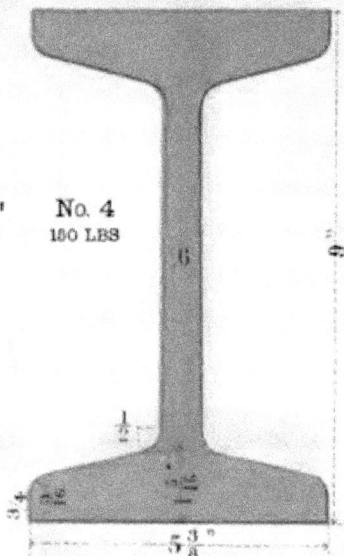

No. 4
150 LBS

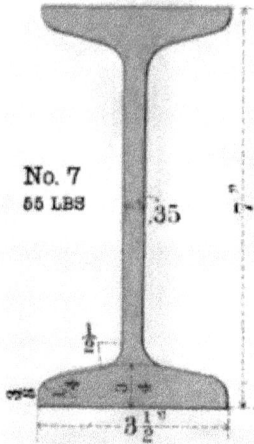

No. 7
55 LBS

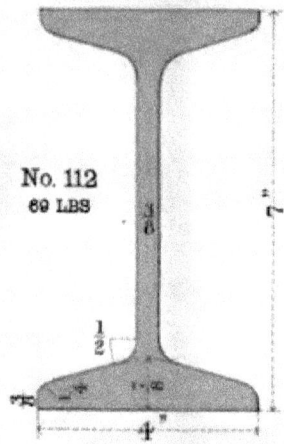

No. 112
69 LBS

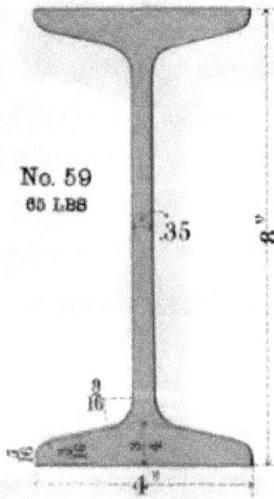

No. 59
65 LBS

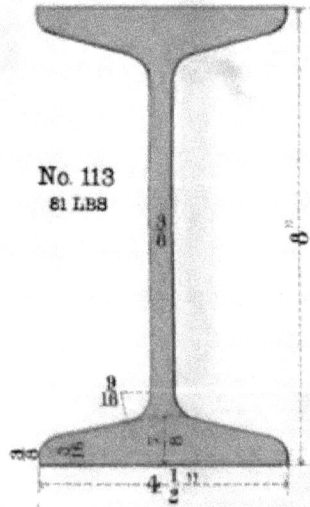

No. 113
81 LBS

NEW BEAMS.

No. 138
125 LBS

No. 139
96 LBS

15"

12"

4 5/8"

4 1/2"

IRON DECK BEAMS.
MINIMUM SIZE.

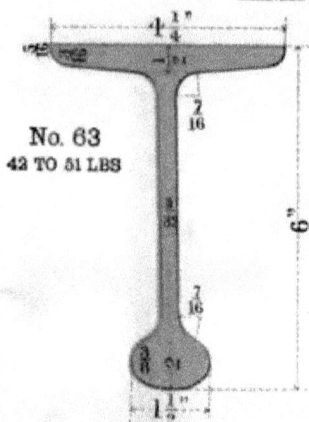

No. 63
42 TO 51 LBS

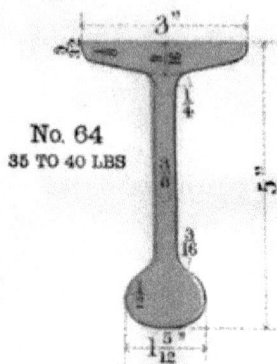

No. 64
35 TO 40 LBS

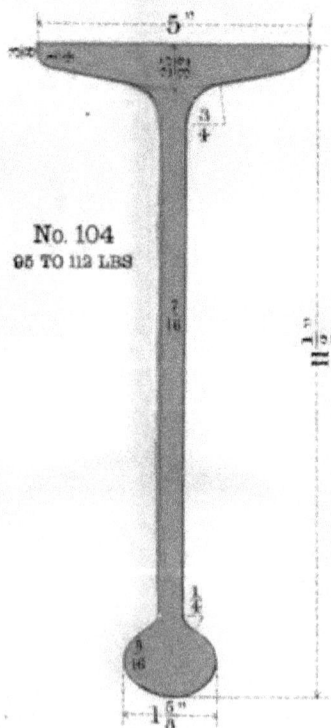

No. 104
95 TO 112 LBS

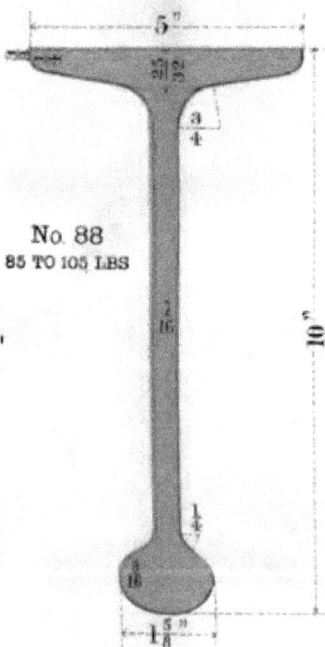

No. 88
85 TO 105 LBS

IRON DECK BEAMS.
MINIMUM SIZE.

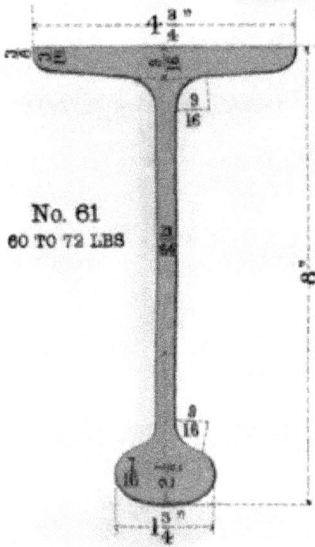

No. 61
60 TO 72 LBS

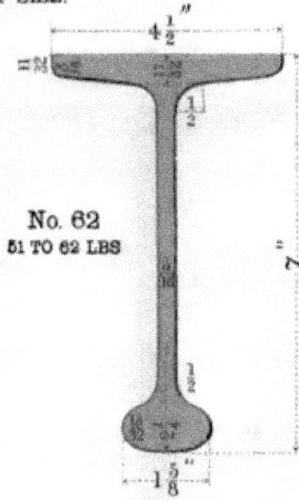

No. 62
51 TO 62 LBS

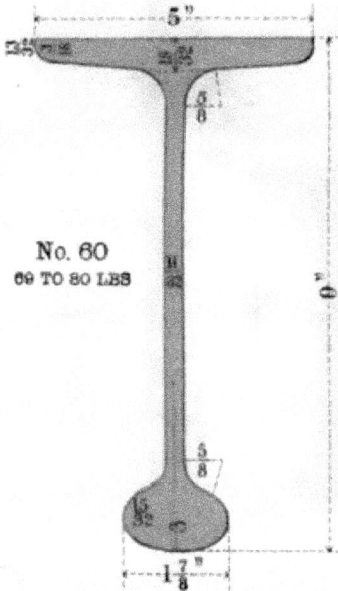

No. 60
69 TO 80 LBS

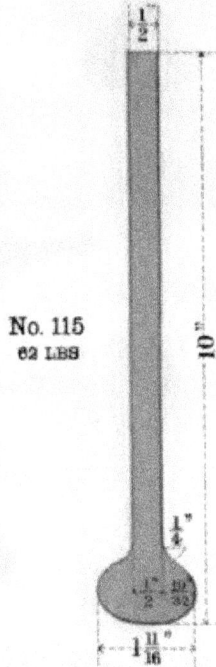

No. 115
62 LBS

STEEL DECK BEAMS.
MINIMUM SIZE.

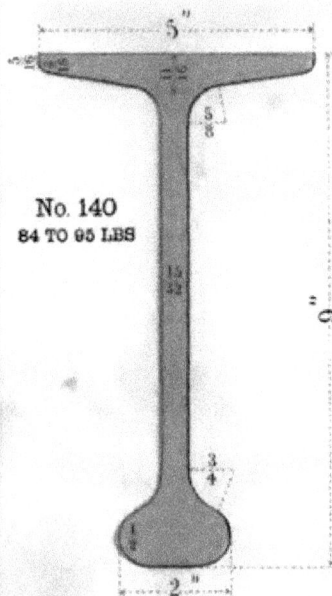

No. 140
84 TO 95 LBS

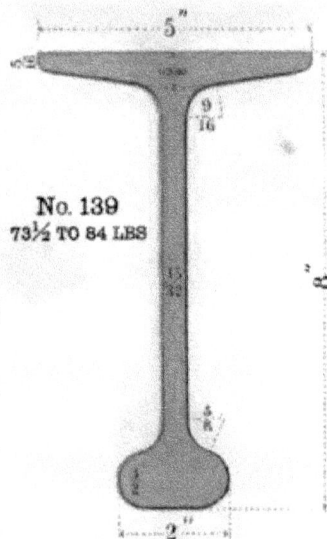

No. 139
73½ TO 84 LBS

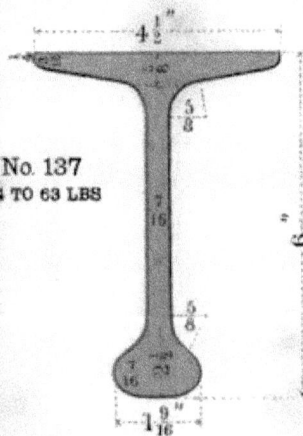

No. 137
54 TO 63 LBS

STEEL DECK BEAMS.
MINIMUM SIZE.

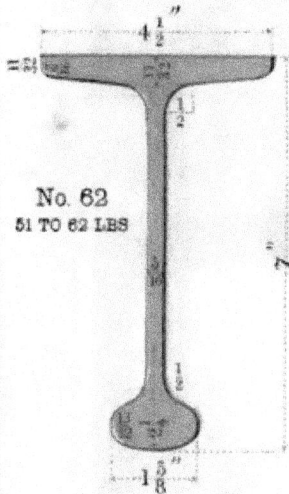

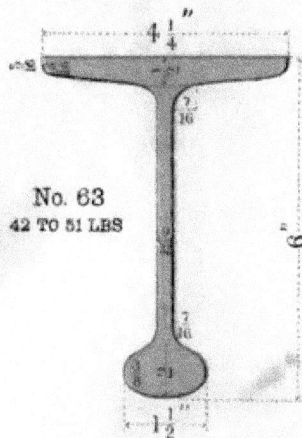

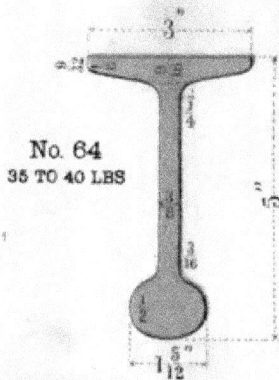

No. 62
51 TO 62 LBS

No. 63
42 TO 51 LBS

No. 64
35 TO 40 LBS

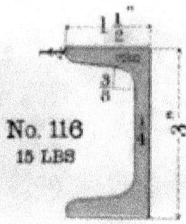

No. 116
15 LBS

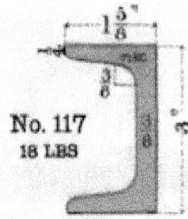

No. 117
18 LBS

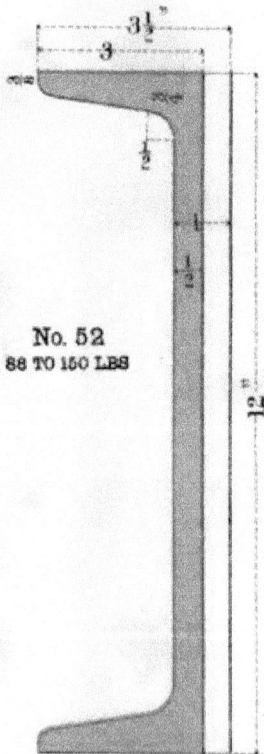

No. 52
88 TO 150 LBS

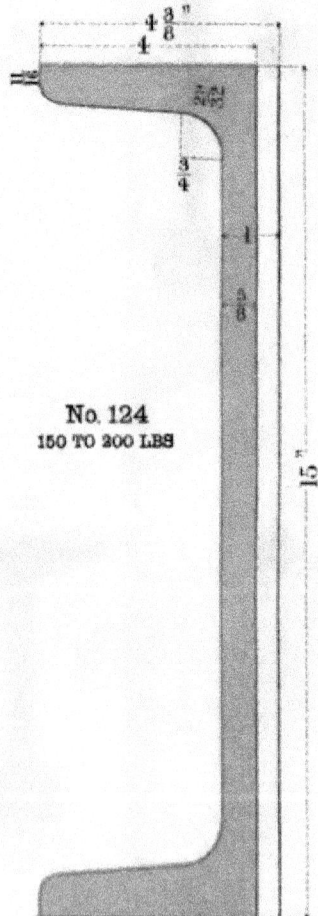

No. 124
150 TO 200 LBS

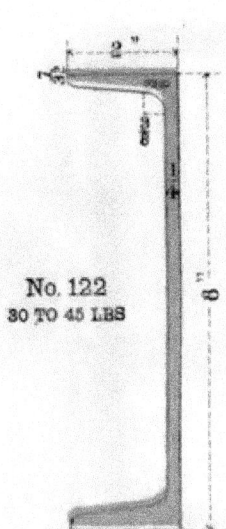

No. 122
30 TO 45 LBS

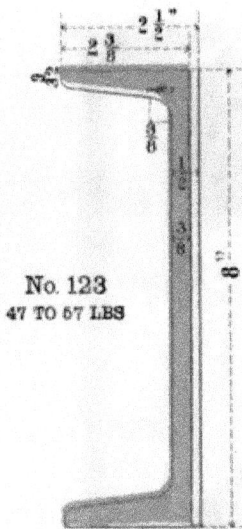

No. 123
47 TO 57 LBS

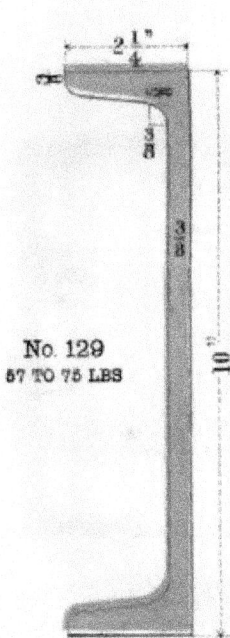

No. 129
57 TO 75 LBS

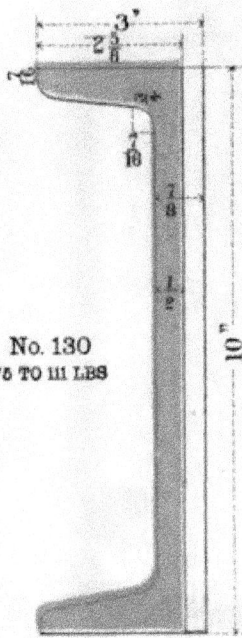

No. 130
75 TO 111 LBS

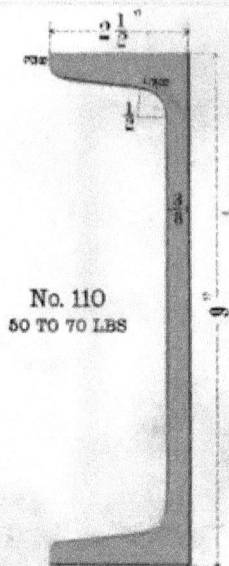

No. 110
50 TO 70 LBS

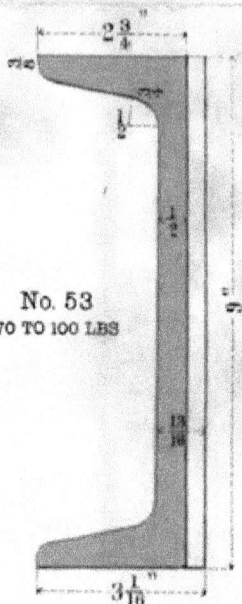

No. 53
70 TO 100 LBS

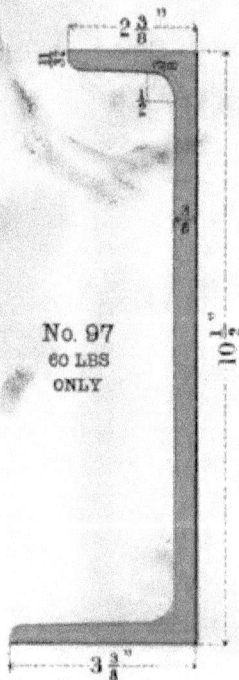

No. 97
60 LBS
ONLY

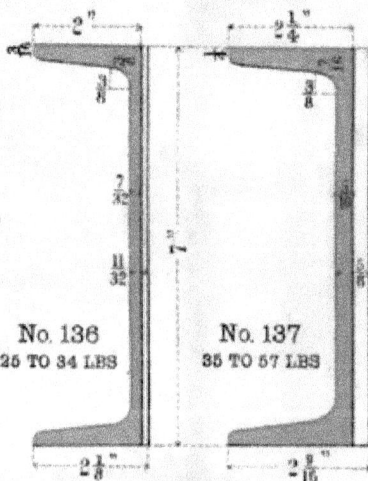

No. 136
25 TO 34 LBS

No. 137
35 TO 57 LBS

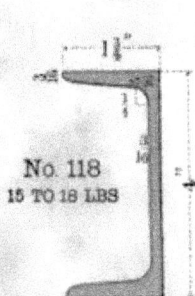

No. 118
15 TO 18 LBS

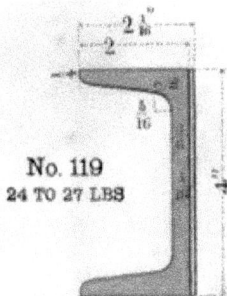

No. 119
24 TO 27 LBS

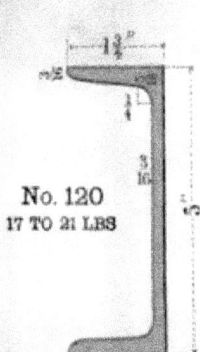

No. 120
17 TO 21 LBS

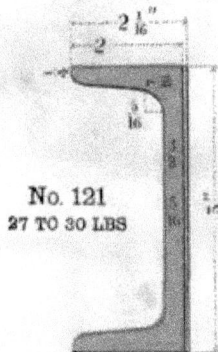

No. 121
27 TO 30 LBS

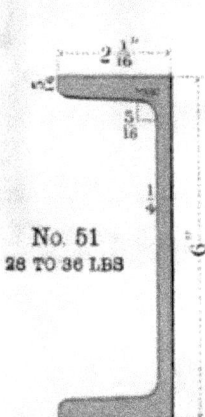

No. 51
28 TO 36 LBS

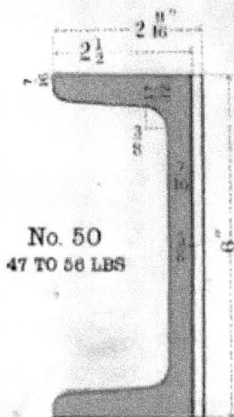

No. 50
47 TO 56 LBS

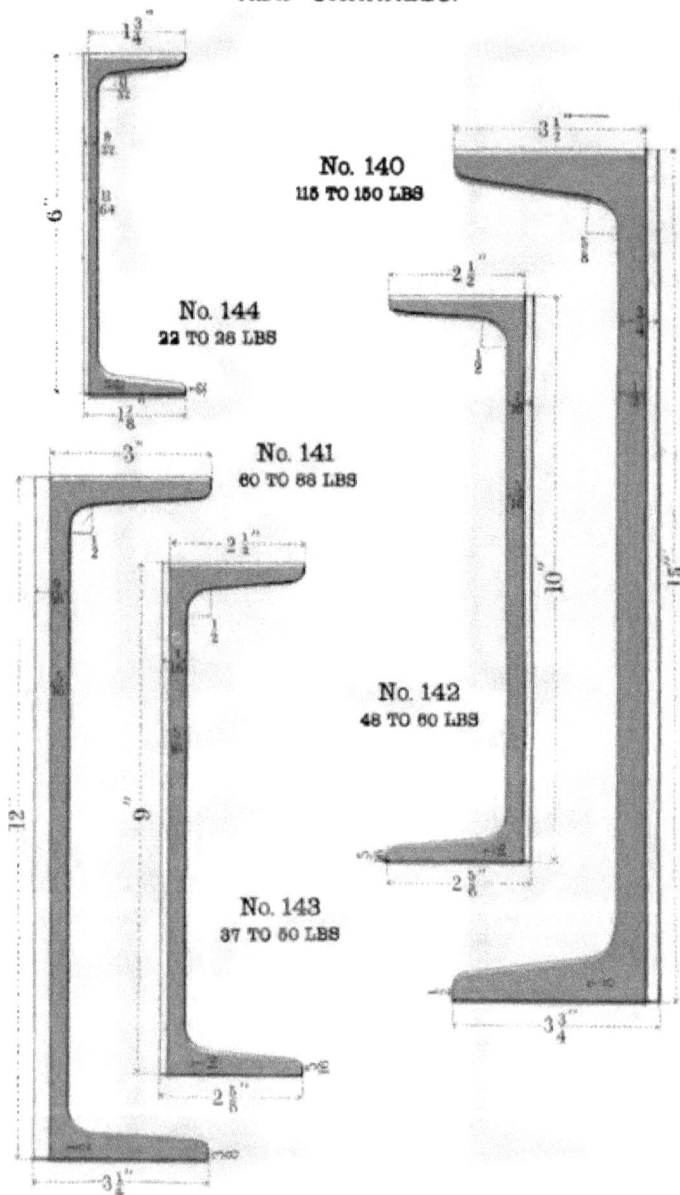

NEW CHANNELS.

No. 140
115 TO 150 LBS

No. 144
22 TO 28 LBS

No. 141
60 TO 88 LBS

No. 142
48 TO 60 LBS

No. 143
37 TO 50 LBS

COLUMN SEGMENTS.

ANY REQUIRED WEIGHT BETWEEN THOSE SPECIFIED WILL BE ROLLED TO ORDER.

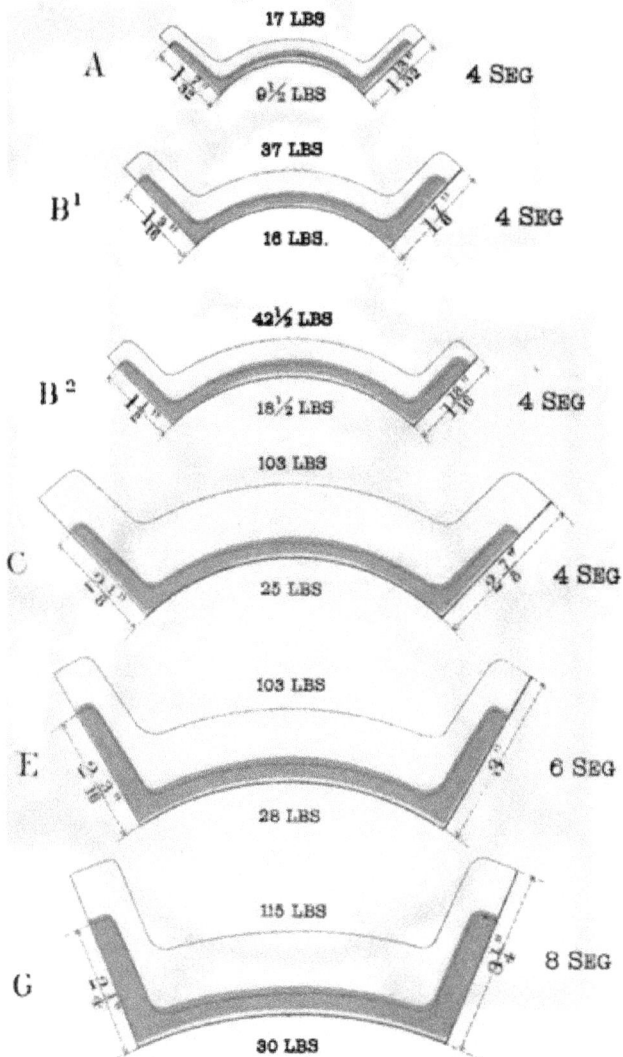

A

17 LBS

9½ LBS

4 SEG

B¹

37 LBS

16 LBS.

4 SEG

B²

42½ LBS

18½ LBS

4 SEG

C

103 LBS

25 LBS

4 SEG

E

103 LBS

28 LBS

6 SEG

G

115 LBS

30 LBS

8 SEG

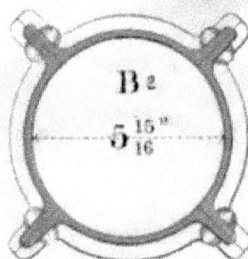

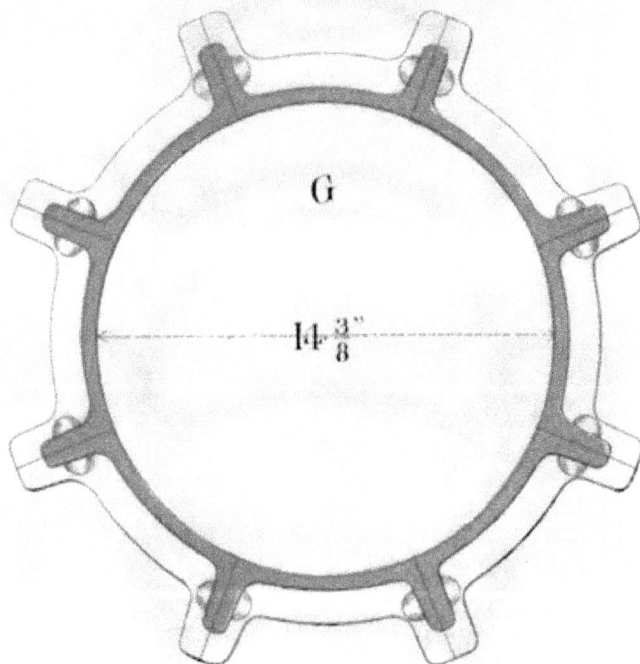

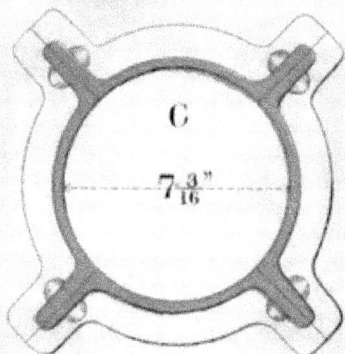

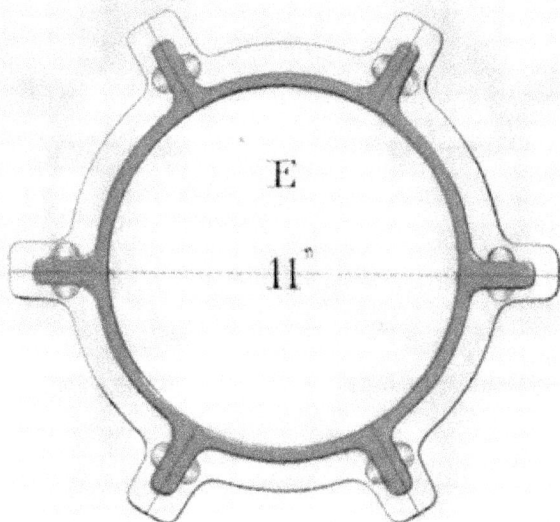

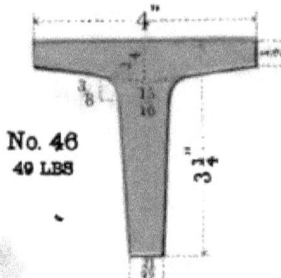

No. 46
49 LBS

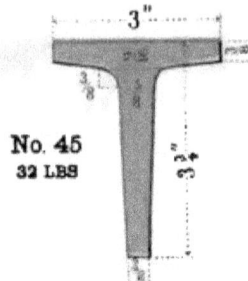

No. 45
32 LBS

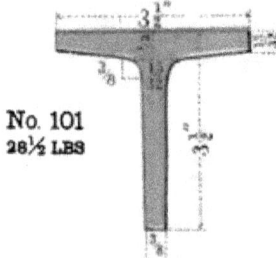

No. 101
28½ LBS

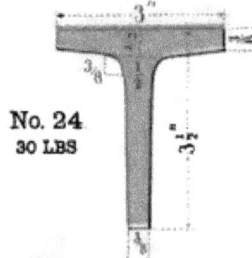

No. 24
30 LBS

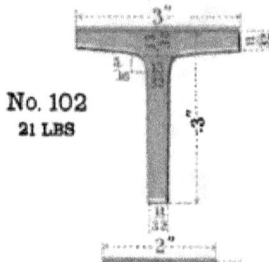

No. 102
21 LBS

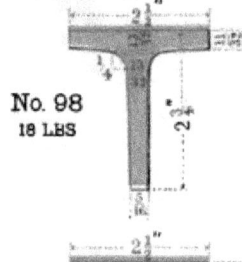

No. 98
18 LBS

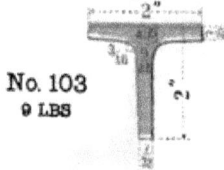

No. 103
9 LBS

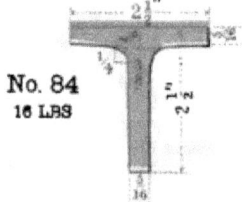

No. 84
16 LBS

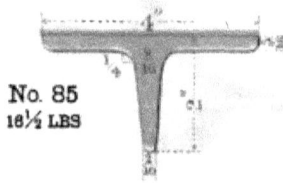

No. 85
16½ LBS

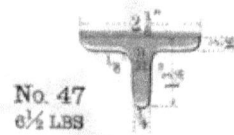

No. 47
6½ LBS

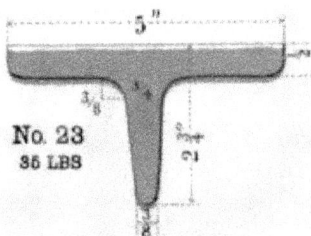

No. 23
35 LBS

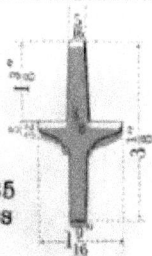

No. 135
14½ LBS

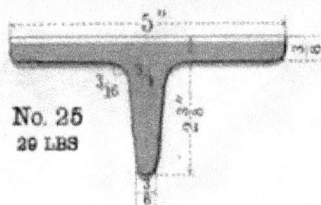

No. 25
29 LBS

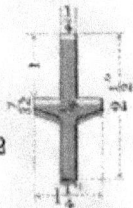

No. 32
9 LBS

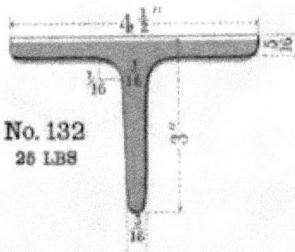

No. 132
25 LBS

No. 34   6 LBS

No. 56   9 LBS

No. 33   4½ LBS

No. 108
15 TO 45 LBS

No. 107
15 TO 45 LBS

EQUAL-SIDED ANGLES.

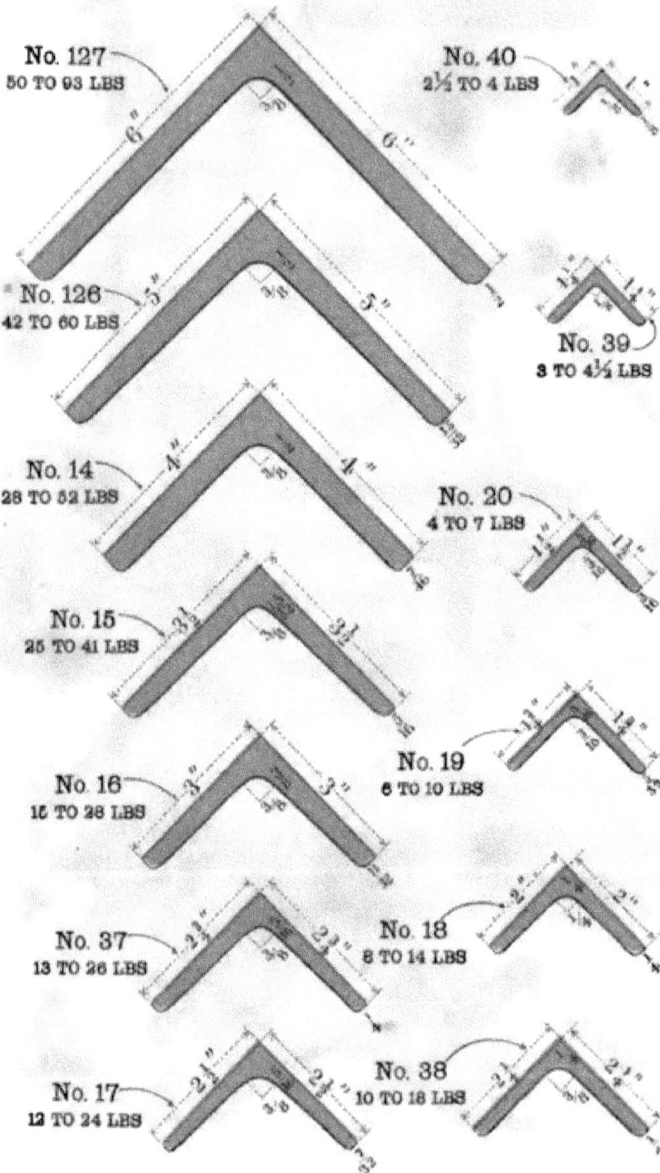

No. 127
50 TO 93 LBS

No. 40
2½ TO 4 LBS

No. 126
42 TO 60 LBS

No. 39
3 TO 4½ LBS

No. 14
28 TO 52 LBS

No. 20
4 TO 7 LBS

No. 15
25 TO 41 LBS

No. 16
15 TO 28 LBS

No. 19
6 TO 10 LBS

No. 37
13 TO 26 LBS

No. 18
8 TO 14 LBS

No. 17
12 TO 24 LBS

No. 38
10 TO 18 LBS

34

UNEQUAL-SIDED ANGLES.

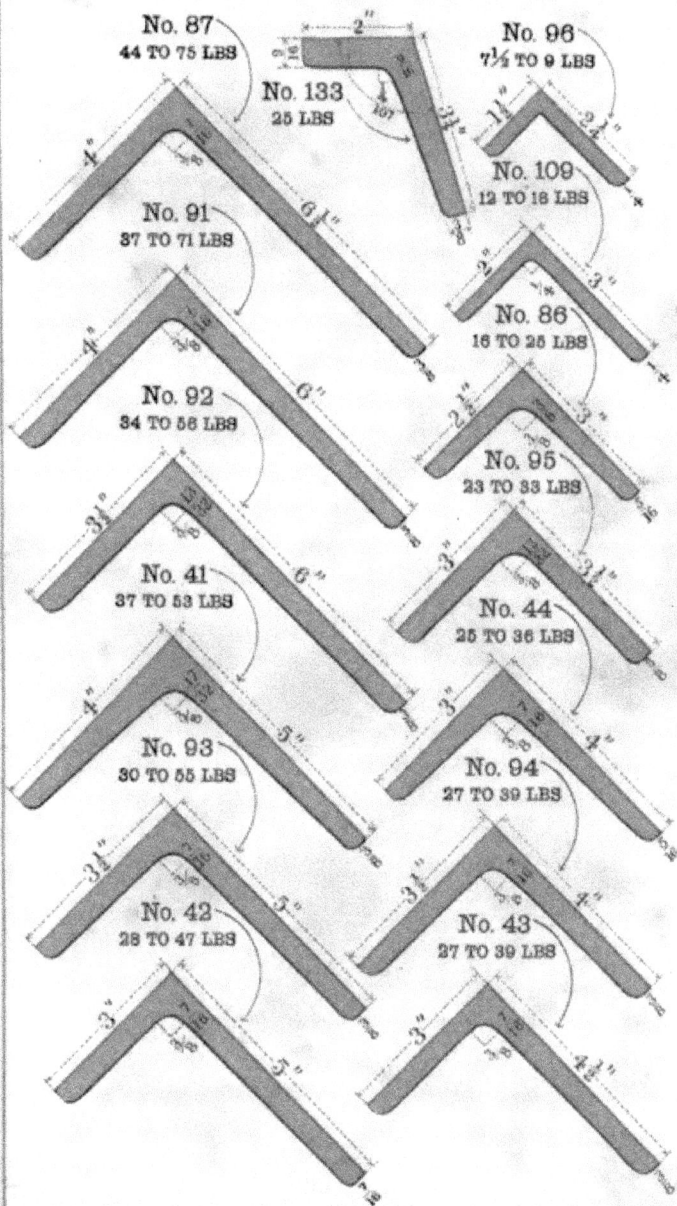

No. 87
44 TO 75 LBS

No. 133
25 LBS

No. 96
7½ TO 9 LBS

No. 91
37 TO 71 LBS

No. 109
12 TO 18 LBS

No. 92
34 TO 56 LBS

No. 86
16 TO 25 LBS

No. 41
37 TO 53 LBS

No. 95
23 TO 33 LBS

No. 93
30 TO 55 LBS

No. 44
25 TO 36 LBS

No. 42
28 TO 47 LBS

No. 94
27 TO 39 LBS

No. 43
27 TO 39 LBS

STANDARD SPACING FOR HOLES
IN BEAM FLANGES.

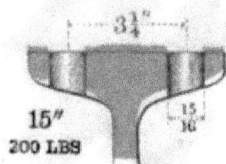

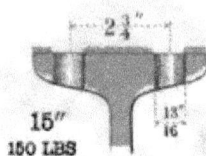

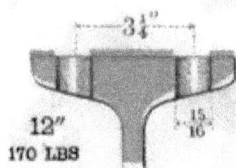

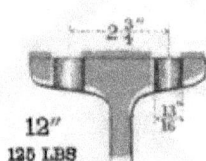

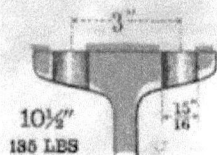

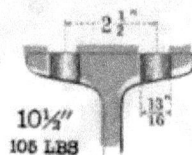

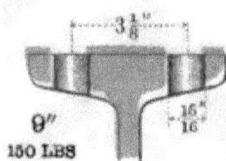

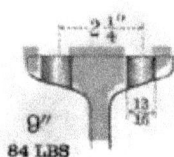

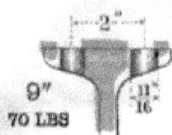

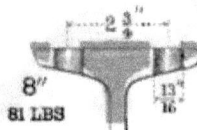

$3\frac{1}{4}''$ **15″** 200 LBS — $\frac{15}{16}''$	$2\frac{1}{4}''$ **15″** 150 LBS — $\frac{13}{16}''$
$3\frac{1}{4}''$ **12″** 170 LBS — $\frac{15}{16}''$	$2\frac{3}{4}''$ **12″** 125 LBS — $\frac{13}{16}''$
$3''$ **10½″** 135 LBS — $\frac{15}{16}''$	$2\frac{1}{2}''$ **10½″** 105 LBS — $\frac{13}{16}''$
$3\frac{1}{8}''$ **9″** 150 LBS — $\frac{16}{16}''$	$2\frac{1}{4}''$ **9″** 84 LBS — $\frac{13}{16}''$
$2''$ **9″** 70 LBS — $\frac{11}{16}''$	$2\frac{3}{4}''$ **8″** 81 LBS — $\frac{13}{16}''$

STANDARD SPACING FOR HOLES
IN BEAM FLANGES.

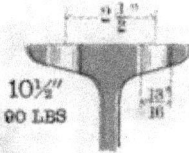

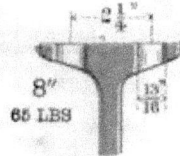

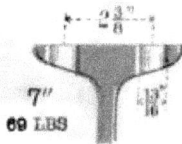

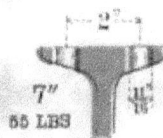

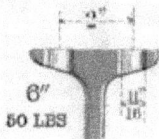

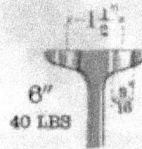

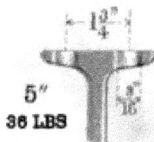

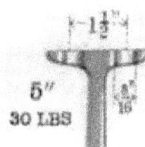

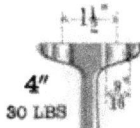

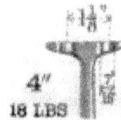

10½" 90 LBS — $2\frac{1}{2}''$ — $\frac{13}{16}''$

8" 65 LBS — $2\frac{1}{4}''$ — $\frac{13}{16}''$

7" 69 LBS — $2\frac{3}{8}''$ — $\frac{13}{16}''$

7" 55 LBS — $2\frac{7}{8}''$ — $\frac{11}{16}''$

6" 50 LBS — $2''$ — $\frac{11}{16}''$

6" 40 LBS — $1\frac{1}{2}''$ — $\frac{9}{16}''$

5" 36 LBS — $1\frac{3}{4}''$ — $\frac{9}{16}''$

5" 30 LBS — $1\frac{1}{2}''$ — $\frac{9}{16}''$

4" 30 LBS — $1\frac{1}{2}''$ — $\frac{9}{16}''$

4" 18 LBS — $1\frac{1}{8}''$ — $\frac{7}{16}''$

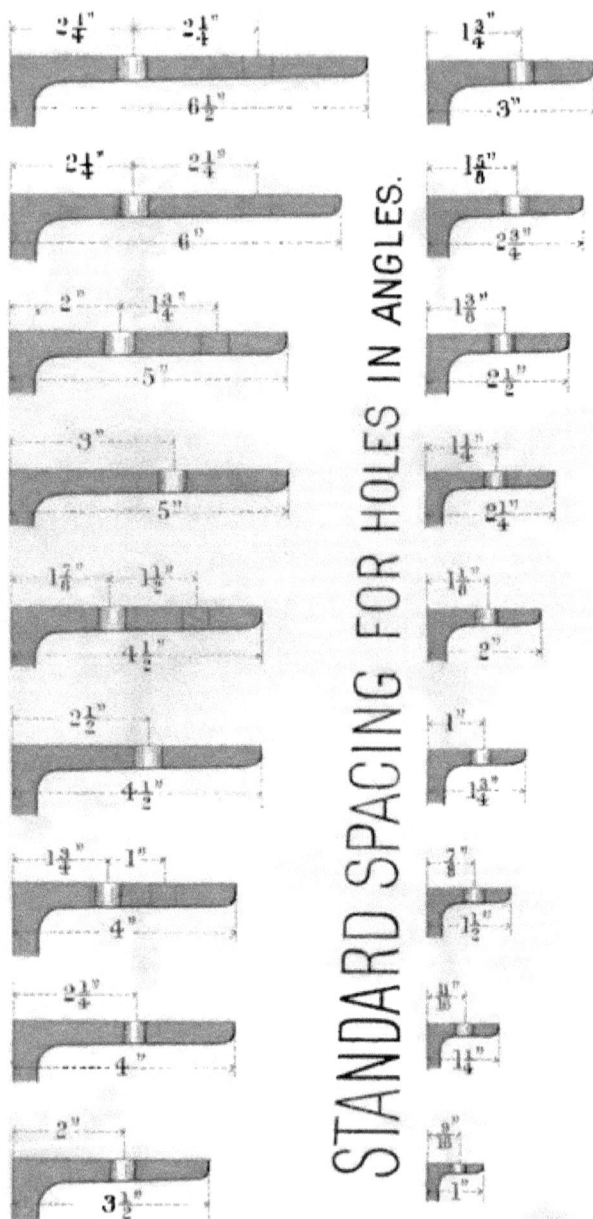

STANDARD SPACING FOR HOLES IN ANGLES.

STANDARD BRACKETS.

FOR 15″ BEAMS

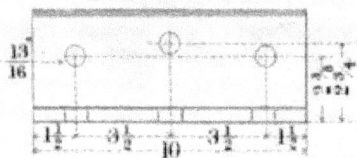
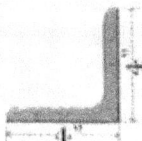

FOR 12″ AND 10½″

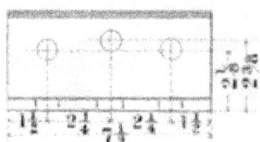
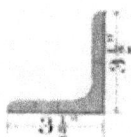

FOR 9″ AND 8″

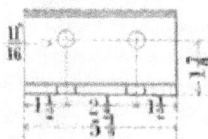

FOR 7″ AND 6″

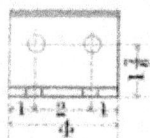
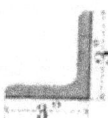

FOR 5″ AND 4″

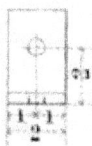
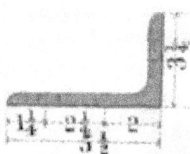

CLEAR SPAN IN FEET.

LOAD PER SQUARE FOOT.

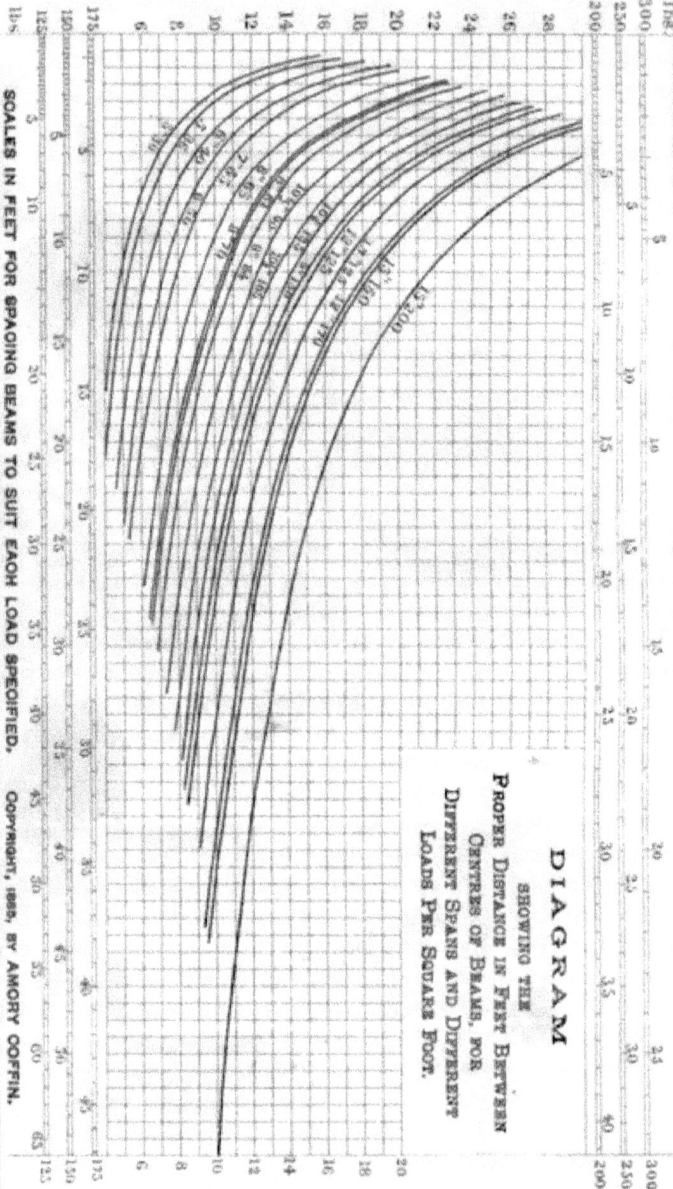

LOAD PER SQUARE FOOT.

SCALES IN FEET FOR SPACING BEAMS TO SUIT EACH LOAD SPECIFIED. COPYRIGHT, 1885, BY AMORY COFFIN.

DIAGRAM

SHOWING THE
PROPER DISTANCE IN FEET BETWEEN
CENTRES OF BEAMS, FOR
DIFFERENT SPANS AND DIFFERENT
LOADS PER SQUARE FOOT.

PRICE CURRENT.

SUBJECT

TO

CHANGES OF MARKET

WITHOUT NOTICE.

NOTE CONCERNING SHAPE IRON.

If any particular dimension is specially desired, attention must be directed to it when ordering, as slight alterations of patterns may occasionally be made in the rolls.

SIZES OF PHŒNIX BAR IRON.

ROUNDS.

$\frac{5}{16}$, $\frac{3}{8}$, $\frac{7}{16}$, $\frac{1}{2}$, $\frac{9}{16}$, $\frac{5}{8}$, $\frac{11}{16}$, $\frac{3}{4}$, $\frac{13}{16}$, $\frac{7}{8}$, $\frac{15}{16}$, 1, $1\frac{1}{8}$, $1\frac{1}{4}$, $1\frac{3}{8}$, $1\frac{1}{2}$, $1\frac{5}{8}$, $1\frac{3}{4}$, $1\frac{7}{8}$, 2, $2\frac{1}{8}$, $2\frac{1}{4}$, $2\frac{3}{8}$, $2\frac{1}{2}$, $2\frac{5}{8}$, $2\frac{3}{4}$, $2\frac{7}{8}$, 3, $3\frac{1}{8}$, $3\frac{1}{4}$, $3\frac{3}{8}$, $3\frac{1}{2}$, $3\frac{5}{8}$, $3\frac{3}{4}$, $3\frac{7}{8}$, 4, $4\frac{1}{4}$, $4\frac{1}{2}$, $4\frac{3}{4}$, 5, $5\frac{1}{4}$, $5\frac{1}{2}$, $5\frac{3}{4}$, 6, $6\frac{1}{4}$, $6\frac{1}{2}$, $6\frac{3}{4}$, 7.

SQUARES.

$\frac{5}{16}$, $\frac{3}{8}$, $\frac{7}{16}$, $\frac{1}{2}$, $\frac{9}{16}$, $\frac{5}{8}$, $\frac{11}{16}$, $\frac{3}{4}$, $\frac{13}{16}$, $\frac{7}{8}$, $\frac{15}{16}$, 1, $1\frac{1}{16}$, $1\frac{1}{8}$, $1\frac{3}{16}$, $1\frac{1}{4}$, $1\frac{3}{8}$, $1\frac{1}{2}$, $1\frac{5}{8}$, $1\frac{3}{4}$, $1\frac{7}{8}$, 2, $2\frac{1}{8}$, $2\frac{1}{4}$, $2\frac{3}{8}$, $2\frac{1}{2}$, $2\frac{5}{8}$, $2\frac{3}{4}$, 3, $3\frac{1}{4}$, $3\frac{1}{2}$, $3\frac{3}{4}$, 4, $4\frac{1}{4}$, $4\frac{1}{2}$, $4\frac{3}{4}$, 5.

FLATS.

Width in Inches.	Thickness in Inches.		Width in Inches.	Thickness in Inches.	
	Min.	Max.		Min.	Max.
$\frac{3}{8}$	$\frac{1}{8}$ to	$\frac{5}{16}$	4	$\frac{1}{4}$ to	$3\frac{1}{2}$
$\frac{15}{16}$	$\frac{1}{16}$ to	$\frac{5}{16}$	$4\frac{1}{4}$	$\frac{1}{4}$ to	$3\frac{1}{2}$
			$4\frac{1}{2}$	$\frac{1}{4}$ to	4
1	$\frac{1}{16}$ to	$\frac{7}{16}$			
$1\frac{1}{8}$	$\frac{1}{16}$ to	$\frac{5}{8}$	5	$\frac{1}{4}$ to	$4\frac{1}{2}$
$1\frac{1}{4}$	$\frac{1}{16}$ to	1	$5\frac{1}{2}$	$\frac{1}{4}$ to	$4\frac{1}{2}$
$1\frac{5}{16}$	$\frac{1}{16}$ to	$1\frac{1}{4}$			
$1\frac{1}{2}$	$\frac{1}{16}$ to	$1\frac{1}{4}$	6	$\frac{1}{4}$ to	5
$1\frac{5}{8}$	$\frac{1}{16}$ to	$1\frac{1}{4}$	$6\frac{1}{2}$	$\frac{1}{4}$ to	2
$1\frac{13}{16}$	$\frac{1}{16}$ to	$1\frac{1}{4}$			
$1\frac{7}{8}$	$\frac{1}{16}$ to	$1\frac{1}{4}$	7	$\frac{1}{4}$ to	$2\frac{1}{2}$
			$7\frac{1}{2}$	$\frac{1}{4}$ to	2
2	$\frac{1}{8}$ to	$1\frac{7}{8}$			
$2\frac{1}{4}$	$\frac{1}{8}$ to	$1\frac{7}{8}$	8	$\frac{1}{4}$ to	$2\frac{1}{2}$
$2\frac{1}{2}$	$\frac{1}{8}$ to	$1\frac{7}{8}$			
$2\frac{3}{4}$	$\frac{1}{8}$ to	$1\frac{7}{8}$	9	$\frac{1}{4}$ to	$1\frac{1}{4}$
			10	$\frac{1}{4}$ to	$1\frac{1}{4}$
3	$\frac{1}{8}$ to	$2\frac{1}{2}$			
$3\frac{1}{4}$	$\frac{1}{4}$ to	$2\frac{3}{4}$	11	$\frac{1}{4}$ to	$1\frac{1}{4}$
$3\frac{1}{2}$	$\frac{1}{4}$ to	3			
$3\frac{3}{4}$	$\frac{1}{4}$ to	$3\frac{1}{4}$	12	$\frac{1}{4}$ to	$1\frac{1}{4}$

ORDINARY SIZES.

$\frac{3}{4}$ to 2 inches. Round and Square . . . . . }

1 to 4 " × $\frac{3}{8}$ to $1\frac{1}{2}$ } Flats . . . . }

$4\frac{1}{8}$ to 6 " × $\frac{3}{8}$ to 1 }

EXTRA SIZES.
● ROUND AND SQUARE. ■

$\frac{5}{16}, \frac{3}{8}, \frac{7}{16}$ . . . . $\frac{1}{10}$c.		$4\frac{1}{8}$ to $4\frac{1}{2}$ . . . . $\frac{6}{10}$c.
$\frac{1}{2}$ and $\frac{9}{16}$ . . . . $\frac{2}{10}$c.		$4\frac{5}{8}$ to 5 . . . . . $\frac{8}{10}$c.
$\frac{5}{8}$ and $\frac{11}{16}$ . . . . $\frac{3}{10}$c.		$5\frac{1}{8}$ to $5\frac{1}{2}$ . . . . 1 c.
$2\frac{1}{8}$ to $2\frac{7}{8}$ . . . . $\frac{1}{10}$c.		$5\frac{3}{4}$ to 6 . . . . $1\frac{5}{10}$c.
3 to $3\frac{1}{2}$ . . . . $\frac{3}{10}$c.		$6\frac{1}{8}$ to $6\frac{1}{2}$ . . . . 2 c.
$3\frac{5}{8}$ to 4 . . . . $\frac{6}{10}$c.		$6\frac{3}{4}$ to 7 . . . . $2\frac{6}{10}$c.

EXTRA SIZES.
FLAT IRON.

$\frac{7}{8} \times \frac{3}{8}$ to $\frac{3}{4}$ . . . $\frac{4}{10}$c.	7 × $2\frac{1}{8}$ to $3\frac{1}{2}$ . . $\frac{6}{10}$c.
1 × $\frac{3}{16}$ . . . . $\frac{4}{10}$c.	$7\frac{1}{2}$ × $\frac{5}{8}$ to 1 . . . $\frac{4}{10}$c.
1 to 6 × $\frac{1}{4}$ and $\frac{5}{16}$ $\frac{2}{10}$c.	$7\frac{1}{2}$ × $1\frac{1}{8}$ to 2 . . . $\frac{4}{10}$c.
2 to 4 × $1\frac{1}{8}$ to 2. $\frac{2}{10}$c.	8 × $\frac{3}{8}$ to 1 . . . $\frac{4}{10}$c.
2 to 4 × $2\frac{1}{8}$ to 3. $\frac{3}{10}$c.	8 × $1\frac{1}{8}$ to $2\frac{3}{4}$ . . $\frac{6}{10}$c.
$4\frac{1}{8}$ to 6 × $1\frac{1}{8}$ to 2. $\frac{2}{10}$c.	9 × $\frac{3}{8}$ to 1 . . . $\frac{6}{10}$c.
$4\frac{1}{8}$ to 6 × $2\frac{1}{8}$ to 3. $\frac{4}{10}$c.	9 × $1\frac{1}{8}$ to 2 . . . $\frac{8}{10}$c.
$6\frac{1}{2}$ × $\frac{3}{8}$ to 1 . . . $\frac{2}{10}$c.	10 × $\frac{3}{8}$ to $1\frac{1}{2}$ . . $\frac{8}{10}$c.
$6\frac{1}{2}$ × $1\frac{1}{8}$ to $2\frac{1}{2}$ . . $\frac{4}{10}$c.	11 × $\frac{3}{8}$ to $1\frac{1}{4}$ . . $\frac{9}{10}$c.
7 × $\frac{3}{8}$ to 1 . . . $\frac{2}{10}$c.	12 × $\frac{5}{8}$ to $1\frac{1}{4}$ . . $\frac{9}{10}$c.
7 × $1\frac{1}{8}$ to 2 . . . $\frac{4}{10}$c.	

$6\frac{1}{2}$ to 12 wide × $\frac{1}{4}$ thick, $\frac{2}{10}$ extra over $\frac{3}{8}$ thick.

ADDITIONAL EXTRAS.
CUTTING TO LENGTHS.
ROUNDS AND SQUARES.

Up to 4 inches, 10 to 20 feet long . . . . $\frac{2}{10}$c.

Over 4 " " " " . . . . $\frac{3}{10}$c.

Under 10 and **over** 20 feet, subject to agreement.

FLATS.

10 to 30 feet long . . . . $\frac{2}{10}$c.

Over 30, for every 10 feet or fraction thereof, $\frac{1}{10}$c. **extra.**

Under 10 feet, subject to agreement.

I BEAMS.

SHAPE.	No.	Depth.	Width of Flange.	Thickness of Web.	Weight per Yard.
		Inches.	*Inches.*	*Inch.*	*Pounds.*
	1	15	5⅜	.65	200
	89	15	4¾	.50	150
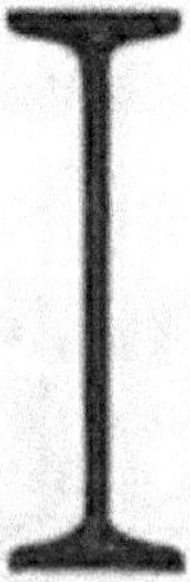	138	15	4⅝	.42	125
	55	12	5½	.59	170
	57	12	4¾	.49	125
	139	12	4½	.38	96
	114	10½	5	.50	135
	58	10½	4½	.44	105
	131	10½	4⅜	.38	90
	4	9	5⅜	.60	150
	5	9	4	.40	84
	6	9	3½	.31	70
	113	8	4½	.38	81
	59	8	4	.35	65
	112	7	4	.38	69
	7	7	3½	.35	55
	111	6	3½	.31	50
	8	6	2¾	.25	40
	106	5	3	.30	36
	105	5	2¾	.25	30
	65	4	2¾	.25	30
	100	4	2	.20	18

To fill special orders, the weight of any of the above can be increased about ten per cent.

DECK BEAMS.

SHAPE.	No.	Depth.	Width of Flange.	Thickness of Web.	Weight per Yard.
		Inches.	Inches.	Inch.	Pounds.
	104	$11\frac{1}{2}$	5	$\frac{7}{16}$	95 to 112
	88	10	5	$\frac{7}{18}$	85 to 105
	60	9	5	$\frac{11}{32}$	69 to 80
	61	8	$4\frac{3}{4}$	$\frac{21}{64}$	60 to 72
	62	7	$4\frac{1}{2}$	$\frac{5}{16}$	51 to 62
	63	6	$4\frac{1}{4}$	$\frac{9}{32}$	42 to 51
	64	5	3	$\frac{1}{4}$	35 to 40

STEEL DECK BEAMS.

	No.	Depth.	Width of Flange.	Thickness of Web.	Weight per Yard.
	140	9	5	$\frac{11}{32}$	84 to 95
	139	8	5	$\frac{11}{32}$	$73\frac{1}{2}$ to 84
	137	6	$4\frac{1}{2}$	$\frac{7}{16}$	54 to 63
	62	7	$4\frac{1}{2}$	$\frac{5}{16}$	51 to 62
	63	6	$4\frac{1}{4}$	$\frac{9}{32}$	42 to 51
	64	5	3	$\frac{1}{4}$	35 to 40

The dimensions given correspond to the minimum weights.

CHANNEL BARS.

SHAPE.	No.	Depth.	Width of Flange.	Thickness of Web.	Weight per Yard.
		Inches.	*Inches.*	*Inch.*	*Pounds.*
	124	15	4	$\frac{5}{8}$	150 to 200
	140	15	$3\frac{1}{2}$	$\frac{1}{2}$	115 to 150
	52	12	3	$\frac{1}{2}$	88 to 150
	141	12	3	$\frac{5}{16}$	60 to 88
	97	$10\frac{1}{2}$	$3\frac{3}{8}$ $2\frac{3}{8}$	$\frac{3}{8}$	60 only
	130	10	$2\frac{5}{8}$	$\frac{1}{2}$	75 to 111
	129	10	$2\frac{1}{4}$	$\frac{3}{8}$	57 to 75
	142	10	$2\frac{1}{2}$	$\frac{5}{16}$	48 to 60
	53	9	$2\frac{3}{4}$	$\frac{1}{2}$	70 to 100
	110	9	$2\frac{1}{2}$	$\frac{3}{8}$	50 to 70
	143	9	$2\frac{1}{2}$	$\frac{5}{16}$	37 to 50
	123	8	$2\frac{3}{8}$	$\frac{3}{8}$	47 to 57
	122	8	2	$\frac{1}{3}$	30 to 45
	137	7	$2\frac{1}{4}$	$\frac{5}{16}$	35 to 57
	136	7	2	$\frac{7}{32}$	25 to 34
	50	6	$2\frac{1}{2}$	$\frac{7}{16}$	47 to 56
	51	6	$2\frac{1}{16}$	$\frac{1}{4}$	28 to 36
	144	6	$1\frac{3}{4}$	$\frac{11}{64}$	22 to 28
	121	5	2	$\frac{5}{16}$	27 to 30
	120	5	$1\frac{3}{4}$	$\frac{3}{16}$	17 to 21
	119	4	2	$\frac{5}{16}$	24 to 27
	118	4	$1\frac{3}{4}$	$\frac{3}{16}$	15 to 18
	117	3	$1\frac{5}{8}$	$\frac{3}{8}$	18 to 21
	116	3	$1\frac{1}{2}$	$\frac{1}{4}$	15 to 18

Any increase in thickness of web adds to the width of flanges and to the weight. No. 97 does not admit of any change in its dimensions. The dimensions given correspond to the minimum weights.

T BARS.

SHAPE.	No.	DIMENSIONS.	Weight per Yard.
		Inches.	*Pounds.*
	23	$5 \times 2\frac{3}{4} \times \frac{1}{2}$	35
	25	$5 \times 2\frac{3}{8} \times \frac{3}{8}$	29
	132	$4\frac{1}{2} \times 3 \times \frac{5}{16}$	25
T	46	$4 \times 3\frac{3}{4} \times \frac{3}{4}$	49
	85	$4 \times 2 \times \frac{5}{16}$	16$\frac{1}{2}$
	101	$3\frac{1}{2} \times 3\frac{1}{2} \times \frac{1}{2}$	28$\frac{1}{2}$
	45	$3 \times 3\frac{3}{4} \times \frac{9}{16}$	32
	24	$3 \times 3\frac{1}{2} \times \frac{1}{2}$	30
	102	$3 \times 3 \times \frac{13}{32}$	21
T	98	$2\frac{1}{2} \times 2\frac{3}{4} \times \frac{13}{32}$	18
	84	$2\frac{1}{2} \times 2\frac{1}{2} \times \frac{3}{8}$	16
	103	$2 \times 2 \times \frac{9}{32}$	9
	47	$2\frac{1}{8} \times 1\frac{5}{16} \times \frac{3}{16}$	6$\frac{1}{2}$

NOTE.—No change can be made in the above dimensions.

EQUAL-SIDED ANGLES.

SHAPE.	No.	DIMENSIONS.	Weight per Yard.
		Inches.	*Pounds.*
	127	$6 \times 6 \times \frac{7}{16}$ to $1\frac{3}{8}$	50.3 to 93.5
	126	$5 \times 5 \times \frac{13}{32}$ to $1\frac{1}{8}$	37.0 to 62.0
	14	$4 \times 4 \times \frac{3}{8}$ to $1\frac{1}{8}$	28.1 to 51.6
	15	$3\frac{1}{2} \times 3\frac{1}{2} \times \frac{5}{16}$ to $\frac{6}{8}$	20.5 to 41.0
	16	$3 \times 3 \times \frac{1}{4}$ to $\frac{1}{2}$	15.0 to 28.1
	37	$2\frac{3}{4} \times 2\frac{3}{4} \times \frac{1}{4}$ to $\frac{1}{2}$	13.4 to 25.8
	17	$2\frac{1}{2} \times 2\frac{1}{2} \times \frac{7}{32}$ to $\frac{1}{2}$	10.5 to 23.6
	38	$2\frac{1}{4} \times 2\frac{1}{4} \times \frac{3}{16}$ to $\frac{7}{16}$	8.0 to 18.3
	18	$2 \times 2 \times \frac{3}{16}$ to $\frac{3}{8}$	7.5 to 14.0
	19	$1\frac{3}{4} \times 1\frac{3}{4} \times \frac{3}{16}$ to $\frac{5}{16}$	6.1 to 10.1
	20	$1\frac{1}{2} \times 1\frac{1}{2} \times \frac{5}{32}$ to $\frac{1}{4}$	4.4 to 7.1
	39	$1\frac{1}{4} \times 1\frac{1}{4} \times \frac{1}{8}$ to $\frac{3}{16}$	2.8 to 4.3
	40	$1 \times 1 \times \frac{1}{8}$ to $\frac{3}{16}$	2.4 to 3.6

NOTE.—The sides of Angles agree only with the *minimum* thickness in table; they increase in width as the thickness increases.

Orders should specify either the thickness or the weight required, but never both.

UNEQUAL-SIDED ANGLES.

SHAPE.	No.	DIMENSIONS.	Weight per Yard.
		Inches.	*Pounds.*
	87	$6\frac{1}{2} \times 4 \times \frac{11}{32}$ to $\frac{3}{4}$	40.7 to 74.8
	91	$6 \times 4 \times \frac{3}{8}$ to $\frac{3}{4}$	36.5 to 71.2
	92	$6 \times 3\frac{1}{2} \times \frac{3}{8}$ to $\frac{5}{8}$	33.8 to 56.2
	41	$5 \times 4 \times \frac{3}{8}$ to $\frac{5}{8}$	31.9 to 53.1
	93	$5 \times 3\frac{1}{2} \times \frac{5}{16}$ to $\frac{11}{16}$	27.5 to 55.0
	42	$5 \times 3 \times \frac{5}{16}$ to $\frac{5}{8}$	23.6 to 47.1
	43	$4\frac{1}{2} \times 3 \times \frac{3}{8}$ to $\frac{9}{16}$	26.5 to 39.7
	94	$4 \times 3\frac{1}{2} \times \frac{3}{8}$ to $\frac{9}{16}$	26.5 to 39.7
	44	$4 \times 3 \times \frac{5}{16}$ to $\frac{9}{16}$	20.5 to 36.9
	95	$3\frac{1}{2} \times 3 \times \frac{5}{16}$ to $\frac{9}{16}$	19.7 to 34.1
	86	$3 \times 2\frac{1}{2} \times \frac{1}{4}$ to $\frac{1}{2}$	13.0 to 25.8
	109	$3 \times 2 \times \frac{1}{4}$ to $\frac{3}{8}$	11.9 to 17.8
	96	$2\frac{1}{4} \times 1\frac{1}{2} \times \frac{3}{16}$ to $\frac{1}{4}$	7.5 to 9.0

See note on opposite page.

PHŒNIX ANGLE IRON.

TABLE OF THICKNESS AND WEIGHT PER YARD,

AS ORDINARILY MADE.

Angle	Size. IN.	Weight LBS.
6 × 6	7/16	50.3
	1/2	57.5
	9/16	64.7
	5/8	71.9
	11/16	79.1
	3/4	86.3
	13/16	93.5
5 × 5	13/32	37.0
	7/16	40.0
	1/2	45.5
	9/16	51.0
	5/8	56.5
	11/16	62.0
4 × 4	3/8	28.1
	7/16	32.8
	1/2	37.5
	9/16	42.2
	5/8	46.9
	11/16	51.6
3½ × 3½	5/16	20.5
	3/8	24.6
	7/16	28.7
	1/2	32.8
	9/16	36.9
	5/8	41.0
3 × 3	1/4	15.0
	5/16	18.2
	3/8	21.5
	7/16	24.8
	1/2	28.1
2¾ × 2¾	1/4	13.4
	5/16	16.5
	3/8	19.6
	7/16	22.7
	1/2	25.8
2½ × 2½	7/32	10.5
	1/4	12.0
	5/16	14.9
	3/8	17.8
	7/16	20.7
	1/2	23.6
2¼ × 2¼	3/16	8.0
	1/4	10.5
	5/16	13.1
	3/8	15.7
	7/16	18.3
2 × 2	3/16	7.5
	7/32	8.5
	1/4	9.4
	5/16	11.7
	3/8	14.0
1¾ × 1¾	3/16	6.1
	1/4	8.1
	5/16	10.1
1½ × 1½	5/32	4.4
	3/16	5.3
	7/32	6.2
	1/4	7.1
1¼ × 1¼	1/8	2.8
	5/32	3.5
	3/16	4.3
1 × 1	1/8	2.4
	5/32	3.0
	3/16	3.6
6½ × 4	13/32	40.7
	7/16	43.8
	1/2	50.0
	9/16	56.2
	5/8	62.4
	11/16	68.6
	3/4	74.8
6 × 4	3/8	36.5
	7/16	41.5
	1/2	47.5
	9/16	53.4
	5/8	59.3
	11/16	65.3
	3/4	71.2
6 × 3½	3/8	33.8
	7/16	39.4
	1/2	45.0
	9/16	50.6
	5/8	56.2
5 × 4	3/8	31.9
	7/16	37.2
	1/2	42.5
	9/16	47.8
	5/8	53.1
5 × 3½	5/16	27.5
	3/8	30.0
	7/16	35.0
	1/2	40.0
	9/16	45.0
	5/8	50.0
	11/16	55.0
5 × 3	5/16	23.6
	3/8	28.3
	7/16	33.0
	1/2	37.7
	9/16	42.4
	5/8	47.1
4½ × 3	3/8	26.5
	7/16	30.9
	1/2	35.3
	9/16	39.7
4 × 3½	3/8	26.5
	7/16	30.9
	1/2	35.3
	9/16	39.7
4 × 3	5/16	20.5
	3/8	24.6
	7/16	28.7
	1/2	32.8
	9/16	36.9
3½ × 3	5/16	19.7
	3/8	23.3
	7/16	26.7
	1/2	30.4
	9/16	34.1
3 × 2½	1/4	13.0
	5/16	16.2
	3/8	19.4
	7/16	22.6
	1/2	25.8
3 × 2	1/4	11.9
	5/16	14.8
	3/8	17.8
2½ × 1½	3/16	7.5
	1/4	9.0

MISCELLANEOUS SHAPES.

SHAPE.	No.	DIMENSIONS.	Weight per Yard.
		Inches.	*Pounds.*
	115	10 $\times$ $\frac{1}{2}$ Bulb	62
	133	$3\frac{1}{4} \times 2 \times \frac{9}{16}$	25
	135	$3\frac{1}{4} \times 1\frac{9}{16} \times \frac{5}{16}$	$14\frac{1}{2}$
	32	$2\frac{1}{2} \times 1\frac{1}{4} \times \frac{1}{4}$	9
	33	$1\frac{3}{4} \times \frac{3}{4} \times \frac{3}{16}$	$4\frac{1}{2}$
	34	$1\frac{3}{4} \times 1\frac{1}{4} \times \frac{3}{16}$	6
	56	$2\frac{1}{4} \times \frac{9}{16}$	9
	107	$7\frac{1}{4} \times \frac{3}{16}$ to $\frac{1}{2}$	15 to 45
	108	Slight difference in shape.	

PRICE OF PHŒNIX COLUMNS.

RIVETED UP AND TURNED OFF AT ENDS TO SPECIFIED
LENGTHS.

ORDINARY LENGTHS.

A columns . . . . 10 to 20 feet.

All other columns . . . . 10 to 30 feet.

Columns longer or shorter than the ordinary lengths will be at an extra price. Any attachments made or work done will increase the cost.

A, B^1, B^2, and C are 4 Segments.
E is 6 Segments. G is 8 Segments.

C, E, and G Columns.

OVER THREE-EIGHTHS OF AN INCH THICK.

Cross section containing over $3\frac{1}{2}$ □ inches per Segment.

ORDINARY SIZES.

10 feet to 30 feet long . . . . . . }   cents per lb.

EXTRAS.
C, E, and G Columns.

OVER THREE-EIGHTHS OF AN INCH THICK.

Cross section containing over $3\frac{1}{2}$ □ inches to each Segment.

Over   30 feet to 40 feet . . . . $\frac{1}{10}$ cent per lb.

    "   40   "   45   "   . . . . $\frac{3}{10}$ "   "

Under 10   "   5   "   . . . . $\frac{3}{10}$ "   "

THREE-EIGHTHS TO ONE-QUARTER.

Cross section containing $3\frac{1}{2}$ □ inches per Segment, or less.

      10 feet to 30 feet . . . . $\frac{2}{10}$ cent per lb.

Over   30   "   40   "   . . . . $\frac{3}{10}$ "   "

    "   40   "   45   "   . . . . $\frac{5}{10}$ "   "

Under 10   "   5   "   . . . . $\frac{8}{10}$ "   "

B² Columns.

OVER THREE-EIGHTHS OF AN INCH THICK.

Cross section containing 10 $\frac{4}{10}$ □ inches, or over.

	10 feet to 30 feet	. . . .	$\frac{1}{10}$	cent per lb.	
Over	30 " 40 "	. . . .	$\frac{2}{10}$	"	"
"	40 " 45 "	. . . .	$\frac{3}{10}$	"	"
Under 10 " 5 "	. . . .	$\frac{2}{10}$	"	"	

THREE-EIGHTHS TO ONE-QUARTER.

Cross section containing 7 $\frac{4}{10}$ □ inches, or over.

	10 feet to 30 feet	. . . .	$\frac{4}{10}$	cent per lb.	
Over	30 " 40 "	. . . .	$\frac{5}{10}$	"	"
Under 10 " 5 "	. . . .	$\frac{4}{10}$	"	"	

B¹ Columns.

OVER THREE-EIGHTHS OF AN INCH THICK.

Cross section containing 9 $\frac{2}{10}$ □ inches, or over.

	10 feet to 30 feet	. . . .	$\frac{3}{10}$	cent per lb.	
Over	30 " 35 "	. . . .	$\frac{5}{10}$	"	"
Under 10 " 5 "	. . . .	$\frac{5}{10}$	"	"

THREE-EIGHTHS TO ONE-QUARTER.

Cross section containing 6 $\frac{2}{10}$ □ inches, or over.

	10 feet to 30 feet	. . . .	$\frac{5}{10}$	cent per lb.	
Over	30 " 35 "	. . . .	$\frac{3}{10}$	"	"
Under 10 " 5 "	. . . .	$\frac{2}{10}$	"	"

A Columns.

THREE-EIGHTHS TO ONE-QUARTER OF AN INCH THICK.

Cross section containing 4 $\frac{2}{10}$ □ inches, or over.

	10 feet to 20 feet	. . . .	1	cent per lb.	
Over	20 " 30 "	. . . .	1 $\frac{2}{10}$	"	"
Under 10 " 5 "	. . . .	1 $\frac{2}{10}$	"	"

UNDER ONE-QUARTER TO THREE-SIXTEENTHS.

Cross section containing 3 $\frac{2}{10}$ □ inches, or over.

	10 feet to 20 feet	. . . .	1 $\frac{2}{10}$	cent per lb.	
Over	20 " 30 "	. . . . . .	1 $\frac{5}{10}$	"	"
Under 10 " 5 "	. . . . . .	1 $\frac{5}{10}$	"	"

LIST OF
DIE-FORGED EYES ON FLAT BARS.

SIZE OF BAR.	Diameter of Pin.	SIZE OF HEAD.	Head Thicker than Bar.	DIE No.
Inches.		*Inches.*		
2 × 5/8	2 1/16	4 × 7/8	1/8	206
2 × 3/4	2 1/16	4 1/2 × 1	1/4	207
2 × 7/8	2 1/16	5 × 1 1/8	1/4	204
2 × 1	2 9/16	5 1/2 × 1 1/2	1/4	205
2 1/2 × 3/4	2 1/16	4 1/2 × 1 1/16	1 5/8	203
2 1/2 × 7/8	2 1/16	5 1/2 × 1		156
2 1/2 × 7/8	2 1/16	6 × 1		77
2 1/2 × 7/8	3 1/16	6 1/4 × 1		160
3 × 5/8	2 11/16	6 × 1		172
3 × 3/4	2 15/16	7 × 1		1
3 × 7/8	3 7/16	7 1/2 × 1 3/8		153
3 × 1	3 7/16	7 1/2 × 1 3/8		152
3 × 1	4 3/16	7 3/4 × 1 1/4		169
3 × 1 1/8	4 3/16	8 1/4 × 1 1/4		144
3 × 1	5 1/16	8 5/8 × 1 5/8		137
3 1/2 × 7/8	2 11/16	7 × 1 1/4		155
3 1/2 × 3/4	3 1/16	7 1/2 × 1 1/2		176
3 1/2 × 1 3/16	3 7/16	8 × 1 7/16		154
3 1/2 × 3/4	3 1/16	8 1/4 × 1 1/4		175
3 1/2 × 3/4	4 7/16	8 1/2 × 1 1/2		157
4 × 1	3	7 1/4 × 1 3/8		159
4 × 1	3 1/16	7 1/2 × 1 3/8		177
4 × 1 1/2	3 7/16	8 1/4 × 1 3/8		150
4 × 1 1/4	3 1/16	8 3/4 × 1 3/8		171
4 × 1	4 3/16	8 3/4 × 1 3/8		167
4 × 1	4 7/16	9 1/4 × 1 3/8		158
4 × 1	4 1/16	9 1/2 × 1 3/8		168
4 × 1 1/16	5 1/16	10 × 1 7/16		97
4 1/2 × 1 1/4	3 7/16	9 × 1 3/8		149
4 1/2 × 1/4	3 1/16	9 1/2 × 1 1/4		170
4 1/2 × 1 1/4	4 1/16	10 × 1 3/8		151
4 1/2 × 1 1/4	5 1/16	10 1/2 × 1 3/8		62
5 × 2	3 11/16	9 1/2 × 2 1/2	1/2	194
5 × 1	4 1/16	10 × 1 1/4	1/2	162
5 × 2	4 1/16	10 × 2 1/2	1/2	161

LIST OF

DIE-FORGED EYES ON FLAT BARS.

SIZE OF BAR.	Diameter of Pin.	SIZE OF HEAD.	Head Thicker than Bar.	DIE No.
Inches.		*Inches.*		
5 × 1	4 11/16	10½ × 1½	½	164
5 × 2	4 11/16	10½ × 2½	½	163
5 × 1⅜	5 3/16	11 × 1⅞	½	91
5 × 1⅝	5 7/16	11½ × 2⅛	½	166
5 × 2	5 11/16	11½ × 2½	½	165
5 × 1¾	6 3/16	12 × 2¼	½	93
5 × 1¾	6 11/16	12½ × 2¼	½	71
6 × 1¾	4 9/16	11 × 2⅜	⅝	178
6 × 2	4 15/16	12 × 2⅝	⅝	173
6 × 2⅜	4 15/16	12 × 3	⅝	174
6 × 1½	6 5/16	13 × 2⅛	⅝	68
6 × 1⅝	6 15/16	14 × 2¼	⅝	179

Dies for flat bars may be used for bars that are thicker or thinner than sizes specified.

The thickness of a bar should never be less than one-fourth of its width nor more than one-half.

UPSET SCREW ENDS ON ROUND BARS.

Diameter of Bars.	Diameter of Upsets.	Length of Upsets.	Threads per Inch.	Diameter of Bars.	Diameter of Upsets.	Length of Upsets.	Threads per Inch.
Inches.	*Inches.*	*Inches*		*Inches.*	*Inches.*	*Inches.*	
⅝	¾	2½	10	1⅞	2¼	7	4
¾	1	2¾	8	2	2⅜	7½	4
⅞	1⅛	3	7	2⅛	2½	8	4
1	1¼	3½	7	2¼	2⅝	8	4
1⅛	1⅜	4	6	2⅜	2¾	8½	3½
1¼	1½	4½	6	2½	2⅞	9	3½
1⅜	1¾	5	5	2⅝	3	9	3½
1½	1⅞	5½	4	2¾	3⅛	9½	3½
1⅝	2	6	4½	2⅞	3⅜	9½	3¼
1¾	2⅛	6½	4½	3	3½	10	3¼

GENERAL FORMULÆ EXPLANATORY OF THE FOLLOWING TABLES AND THEIR APPLICATION.

Let A represent the area of cross section in square inches.

Let I represent the moment of inertia of A about an axis passing through its centre of gravity.

Let d represent the distance, in inches, of the most remote fibre from the axis for I.

Let $r = \left(\frac{I}{A}\right)^{\frac{1}{2}}$ represent the radius of gyration of the section A.

All the preceding quantities are given in the following tables for the various sections of beams, channels, angles, etc.

Let M represent the greatest bending moment, in inch-pounds, for any loading or span.

Let l represent the span in feet.

With the load W *pounds at the centre of the span l* :—

$M = 3\,W\,l$ for ends of beam simply supported.

$M = \left\{ \begin{array}{c} \frac{16}{8}\,W\,l \\ -\frac{2}{4}\,W\,l \end{array} \right\}$ for one end simply supported and the other fixed.

$M = \left\{ \begin{array}{c} \frac{3}{8}\,W\,l \\ -\frac{3}{2}\,W\,l \end{array} \right\}$ for both ends of beam fixed.

With the uniform load of w pounds per lineal foot of span :—

$M = \frac{3}{2}\,w\,l^2$ for ends of beam simply supported.

$M = \left\{ \begin{array}{c} \frac{27}{7}\,w\,l^2 \\ -\frac{3}{2}\,w\,l^2 \end{array} \right\}$ for one end simply supported and the other fixed.

$M = \left\{ \begin{array}{c} \frac{1}{2}\,w\,l^2 \\ -\,w\,l^2 \end{array} \right\}$ for both ends of beam fixed.

The preceding negative values belong to points of support.

Let K represent the greatest stress in pounds per square inch,—*i.e.*, the stress in the most remote fibre.

Then $M = \dfrac{K I}{d}$ . . . . (1):

Or, $K = \dfrac{M d}{I}$ . . . . (2).

If r is known, as it sometimes may be,

$$A = \dfrac{M d}{K r^2} .$$ . . . . . . (3).

Let D represent the greatest deflection in inches.

Let E represent the coefficient of elasticity in pounds per square inch. Then

W at span centre.   Uniform load.

$D = \quad 36 \dfrac{W \, l^3}{E \, I}$ ...... $22.5 \dfrac{w \, l^4}{E \, I}$ for supported ends.

$D = 17.11 \dfrac{W \, l^3}{E \, I}$ ......$9.366 \dfrac{w \, l^4}{E \, I}$ for one supported and one fixed end.

$D = \quad 9 \dfrac{W \, l^3}{E \, I}$ ...... $4.5 \dfrac{w \, l^4}{E \, I}$ for both ends fixed.

For a circular section $I = \dfrac{\pi R^4}{4}$ and $d = R$ (the radius).

Hence, $M = 0.7854 \, K \, R^3$ . . . . . (4).

Eqs. (1), (2), (3), and (4) are of great practical value. The values in table on page 58 are computed from Eq. (4), with K equal to 15,000, 18,000, and 20,000.

RIVET BEARING AND SHEARING.

Let S represent the shearing resistance in pounds per square inch.

Let p represent the bearing pressure in pounds per square inch.

Let (2R) represent the rivet diameter in inches.

Let t represent the thickness of plate in inches.

Then, Shearing resistance of rivet $= \pi R^2 S$. (5).

Bearing resistance of rivet $= 2R \, pt$. (6).

The values of Eqs. (5) and (6) for $S = 7500$, and $p = 12,000$ and $15,000$ are given for various values of (2R) and t on page 59.

MAXIMUM BENDING MOMENTS TO BE ALLOWED ON PINS FOR FIBRE STRAINS OF 15,000, 18,000, AND 20,000 POUNDS.

Diam. of Pin. Inches.	BENDING MOMENTS.			Diam. of Pin. Inches.	BENDING MOMENTS.		
	S=15,000	S=18,000	S=20,000		S=15,000	S=18,000	S=20,000
1	1,470	1,770	1,960	3 9/16	66,580	79,900	88,770
1 1/16	1,770	2,120	2,350	3 5/8	70,140	84,170	93,520
1 1/8	2,100	2,520	2,800	3 11/16	73,840	88,600	98,450
1 3/16	2,470	2,960	3,290	3 3/4	77,660	93,190	103,550
1 1/4	2,880	3,450	3,830	3 13/16	81,600	97,920	108,800
1 5/16	3,330	4,000	4,440	3 7/8	85,690	102,820	114,250
1 3/8	3,830	4,590	5,100	3 15/16	89,900	107,880	119,870
1 7/16	4,370	5,250	5,830	4	94,240	113,090	125,660
1 1/2	4,970	5,960	6,630	4 1/16	98,720	118,460	131,620
1 9/16	5,620	6,740	7,490	4 1/8	103,370	124,040	137,820
1 5/8	6,320	7,580	8,420	4 3/16	108,130	129,760	144,170
1 11/16	7,080	8,490	9,430	4 1/4	113,040	135,650	150,720
1 3/4	7,890	9,470	10,520	4 5/16	118,100	141,730	157,470
1 13/16	8,770	10,520	11,690	4 3/8	123,320	147,980	164,420
1 7/8	9,710	11,650	12,940	4 7/16	128,680	154,420	171,570
1 15/16	10,710	12,850	14,280	4 1/2	134,190	161,030	178,920
2	11,780	14,140	15,710	4 9/16	139,860	167,830	186,480
2 1/16	12,920	15,500	17,220	4 5/8	145,690	174,820	194,250
2 1/8	14,130	16,960	18,840	4 11/16	151,670	182,000	202,220
2 3/16	15,410	18,500	20,550	4 3/4	157,820	189,380	210,450
2 1/4	16,770	20,130	22,360	4 13/16	164,140	196,960	218,850
2 5/16	18,210	21,850	24,280	4 7/8	170,600	204,750	227,470
2 3/8	19,720	23,670	26,300	4 15/16	177,260	212,710	236,350
2 7/16	21,320	25,590	28,430	5	184,100	220,800	245,400
2 1/2	23,000	27,600	30,670	5 1/8	198,200	237,800	264,300
2 9/16	24,780	29,730	33,040	5 1/4	213,100	255,600	284,100
2 5/8	26,620	31,950	35,500	5 3/8	228,700	274,300	304,900
2 11/16	28,580	34,300	38,110	5 1/2	245,000	294,000	326,700
2 3/4	30,630	36,750	40,830	5 5/8	262,100	314,400	349,500
2 13/16	32,760	39,310	43,680	5 3/4	280,000	336,000	373,300
2 7/8	34,980	41,980	46,650	5 7/8	298,600	358,300	398,200
2 15/16	37,330	44,800	49,770	6	318,100	381,700	424,100
3	39,750	47,700	53,000	6 1/8	338,400	406,100	451,200
3 1/16	42,290	50,750	56,390	6 1/4	359,500	431,400	479,400
3 1/8	44,940	53,930	59,920	6 3/8	381,500	457,830	508,700
3 3/16	47,690	57,230	63,590	6 1/2	404,400	485,300	539,200
3 1/4	50,550	60,660	67,400	6 5/8	428,200	513,900	570,900
3 5/16	53,520	64,230	71,370	6 3/4	452,900	543,300	603,900
3 3/8	56,600	67,930	75,470	6 7/8	478,500	574,200	638,000
3 7/16	59,810	71,780	79,750	7	505,100	606,100	673,500
3 1/2	63,130	75,760	84,180				

SHEARING AND BEARING VALUES OF RIVETS.

Diam. of Rivet	Single Shear at 7500 Lbs	Allowed Bearing Pressure Per Sq. In.	BEARING VALUE FOR DIFFERENT THICKNESSES OF PLATE												
			1/4"	5/16"	3/8"	7/16"	1/2"	9/16"	5/8"	11/16"	3/4"	13/16"	7/8"	15/16"	1"
3/8"	828	15,000	1,410												
		12,000	1,120	1,410											
7/16"	1,130	15,000	1,640	2,050											
		12,000	1,310	1,640	1,970										
1/2"	1,470	15,000	1,870	2,340	2,810										
		12,000	1,500	1,870	2,250	2,620									
9/16"	1,860	15,000	2,110	2,640	3,160	3,690									
		12,000	1,690	2,110	2,530	2,950	3,370								
5/8"	2,300	15,000	2,340	2,930	3,520	4,100									
		12,000	1,870	2,340	2,810	3,280	3,750	4,220							
11/16"	2,780	15,000	2,580	3,220	3,870	4,510	5,160								
		12,000	2,060	2,580	3,090	3,610	4,120	4,640	5,160						
3/4"	3,310	15,000	2,810	3,520	4,220	4,920	5,620	6,330							
		12,000	2,250	2,810	3,370	3,940	4,500	5,060	5,620	6,190					
13/16"	3,890	15,000	3,050	3,810	4,570	5,330	6,090	6,860	7,620						
		12,000	2,440	3,050	3,660	4,270	4,870	5,480	6,090	6,700	7,310				
7/8"	4,510	15,000	3,280	4,100	4,920	5,740	6,560	7,380	8,200	9,020					
		12,000	2,620	3,280	3,940	4,590	5,250	5,910	6,560	7,220	7,870	8,530			
15/16"	5,180	15,000	3,520	4,390	5,270	6,150	7,030	7,910	8,790	9,670					
		12,000	2,810	3,520	4,220	4,920	5,620	6,330	7,030	7,730	8,440	9,140	9,840		
1"	5,890	15,000	3,750	4,690	5,620	6,560	7,500	8,440	9,370	10,310	11,250				
		12,000	3,000	3,750	4,500	5,250	6,000	6,750	7,500	8,250	9,000	9,750	10,500	11,250	
1 1/16"	6,650	15,000	3,980	4,980	5,980	6,970	7,970	8,960	9,960	10,960	11,950	12,950			
		12,000	3,190	3,980	4,780	5,580	6,370	7,170	7,970	8,770	9,560	10,360	11,160	11,950	12,750
1 1/8"	7,460	15,000	4,220	5,270	6,330	7,380	8,440	9,490	10,550	11,600	12,660	13,710	14,770		
		12,000	3,370	4,220	5,060	5,910	6,750	7,590	8,440	9,280	10,120	10,970	11,810	12,650	13,500
1 3/16"	8,310	15,000	4,450	5,570	6,680	7,790	8,910	10,020	11,130	12,250	13,360	14,470	15,590		
		12,000	3,560	4,450	5,340	6,230	7,120	8,010	8,910	9,800	10,690	11,580	12,470	13,360	14,250

PROPERTIES OF PHŒNIX BEAMS.

No. of Shape.	DESIGNATION.	Weight Per Yard. lbs.	Area of Section. Sq. In.	Thickness of Web. Inches.	Width of Flange. Inches.	MOMENT OF INERTIA.		RADIUS OF GYRATION.	
						Neutral Axis Perpendicular to Axis of Web.	Neutral Axis Coincident with Axis of Web.	Neutral Axis Perpendicular to Axis of Web.	Neutral Axis Coincident with Axis of Web.
1	15″ Heavy.	200	20.0	0.65	5.38	676.57	23.93	5.82	1.09
89	15″ Medium.	150	15.0	0.5	4.75	506.74	13.62	5.81	0.95
138	15″ Light.	125	12.5	0.42	4.63	416.19	10.26	5.77	0.90
55	12″ Heavy.	170	17.0	0.59	5.5	381.91	24.08	4.74	1.19
57	12″ Medium.	125	12.5	0.49	4.75	282.56	12.98	4.75	1.02
139	12″ Light.	96	9.6	0.375	4.5	201.65	7.60	4.58	0.89
114	10½″ Heavy.	135	13.5	0.5	5.0	240.59	16.72	4.23	1.11
58	10½″ Medium.	105	10.5	0.44	4.5	175.36	9.03	4.09	0.93
131	10½″ Light.	90	9.0	0.375	4.38	158.68	7.63	4.20	0.92
4	9″ Heavy.	150	15.0	0.6	5.38	189.07	23.16	3.55	1.24
5	9″ Medium.	84	8.4	0.4	4.0	110.93	6.28	3.63	0.86
6	9″ Light.	70	7.0	0.31	3.5	86.97	3.62	3.53	0.72
113	8″ Heavy.	81	8.1	0.375	4.5	84.44	7.69	3.23	0.98
59	8″ Light.	65	6.5	0.38	4.0	68.54	4.58	3.25	0.84
112	7″ Heavy.	69	6.9	0.375	4.0	55.74	5.42	2.84	0.89
7	7″ Light.	55	5.5	0.35	3.5	44.22	3.27	2.83	0.77
111	6″ Heavy.	50	5.0	0.31	3.5	29.65	2.79	2.43	0.75
8	6″ Light.	40	4.0	0.25	2.75	21.69	1.25	2.33	0.56
106	5″ Heavy.	36	3.6	0.3	3.0	14.91	1.74	2.04	0.70
105	5″ Light.	30	3.0	0.25	2.75	12.42	1.11	2.03	0.61
65	4″ Heavy.	30	3.0	0.25	2.75	7.63	1.13	1.59	0.61
100	4″ Light.	18	1.8	0.2	2.0	4.41	0.31	1.56	0.42

PROPERTIES OF PHŒNIX DECK BEAMS.

IRON.

No. of Shape.	Designation.	Weight Per Yard Lbs.	Area, Sq. In.	Thickness of Web, Inches.	Width of Flange, Inches.	Moment of Inertia.		Radius of Gyration.		Distance of Centre of Gravity from Outside of Flange.
						Neutral Axis Parallel to Flange.	Neutral Axis Coincident with Web Axis.	Neutral Axis Parallel to Flange.	Neutral Axis Coincident with Web Axis.	
104	11½″ Light.	95	9.5	0.438	5.0	168.75	5.17	4.21	0.74	4.27
88	10″ Light.	85	8.5	0.438	5.0	151.53	5.16	4.22	0.78	3.77
60	9″ Light.	69	6.9	0.344	5.0	73.69	4.84	3.27	0.84	2.96
61	8″ Light.	60	6.0	0.328	4.75	50.37	3.85	2.90	0.80	2.59
62	7″ Light.	51	5.1	0.313	4.5	32.58	3.04	2.53	0.77	2.25
63	6″ Light.	42	4.2	0.281	4.25	19.69	2.35	2.17	0.75	1.88
64	5″ Light.	35	3.5	0.375	3.0	11.27	0.89	1.79	0.51	2.41
STEEL.										
140	9″ Light.	84	8.23	0.438	5.0	101.08	4.55	3.50	0.74	3.59
139	8″ Light.	73.5	7.2	0.438	5.0	66.67	4.55	3.04	0.79	3.40
138	7″ Light.	75	7.35	0.438	5.0	53.13	5.34	2.69	0.85	2.85
137	6″ Heavy.	54	5.3	0.438	4.5	25.16	3.08	2.18	0.76	2.39
136	6″ Light.	42	4.11	0.375	4.0	19.09	2.15	2.16	0.72	2.78

PROPERTIES OF PHŒNIX CHANNEL IRON.

No. of Shape.	DESIGNATION.	Weight Per Yard.	Area of Section. Sq. In.	Width of Flange. Inches.	Thickness of Web. Inches.	MOMENT OF INERTIA.		RADIUS OF GYRATION.		Distance of Centre of Gravity from Outside of Web.
						Neutral Axis Perpendicular to Web Axis at Centre.	Neutral Axis Parallel to Web through Centre of Gravity.	Neutral Axis Perpendicular to Web Axis at Centre.	Neutral Axis Parallel to Web through Centre of Gravity.	
124	15" Heavy.	200	20.0	4.38	1.0	554.57	23.61	5.27	1.09	1.08
124	15" Light.	150	15.0	4.0	0.625	449.11	18.27	5.47	1.10	1.00
140	15" Heavy.	150	15.0	3.75	0.75	421.87	12.39	5.30	0.91	0.86
140	15" Light.	115	11.5	3.50	0.5	351.56	10.01	5.53	0.93	0.83
52	12" Heavy.	150	15.0	3.5	1.0	235.73	8.44	3.96	0.75	0.80
52	12" Light.	88	8.8	3.0	0.5	163.73	5.07	4.31	0.76	0.69
141	12" Heavy.	88	8.8	3.25	0.563	159.44	4.19	4.26	0.69	0.82
141	12" Light.	60	6.0	3.0	0.313	123.50	3.01	4.54	0.71	0.86
130	10" Heavy.	111	11.1	3.0	0.875	128.61	5.26	3.40	0.69	0.76
130	10" Light.	75	7.5	2.63	0.5	97.36	3.51	3.60	0.69	0.66
142	10" Heavy.	60	6.0	2.63	0.438	74.09	2.59	3.51	0.66	0.56
142	10" Light.	48	4.8	2.5	0.313	63.67	2.21	3.64	0.68	0.56
129	10" Heavy.	75	7.5	2.43	0.555	88.17	2.49	3.43	0.58	0.56
129	10" Light.	57	5.7	2.25	0.375	73.17	1.97	3.58	0.59	0.53
53	9" Heavy.	100	10.0	3.06	0.813	94.27	5.24	3.07	0.72	0.76
53	9" Light.	70	7.0	2.75	0.5	75.29	3.69	3.28	0.73	0.70

110	9'' Heavy.	70	7.0	2.72	0.597	74.49	3.18	3.26	0.67	0.71
110	9'' Light.	50	5.0	2.5	0.375	61.01	2.36	3.49	0.69	0.70
143	9'' Heavy.	50	5.0	2.63	0.438	56.83	2.30	3.37	0.68	0.65
143	9'' Light.	37	3.7	2.5	0.313	49.23	1.82	3.65	0.70	0.72
123	8'' Heavy.	57	5.7	2.5	0.5	43.99	2.14	2.76	0.61	0.56
123	8'' Light.	47	4.7	2.38	0.375	38.65	1.82	2.87	0.62	0.55
122	8'' Heavy.	45	4.5	2.2	0.45	34.74	1.14	2.78	0.50	0.47
122	8'' Light.	30	3.0	2.0	0.25	26.20	0.85	2.96	0.53	0.45
137	7'' Heavy.	57	5.7	2.56	0.625	32.69	2.00	2.40	0.59	0.59
137	7'' Light.	35	3.5	2.25	0.313	23.76	1.31	2.61	0.61	0.55
136	7'' Heavy.	34	3.4	2.13	0.344	21.20	0.94	2.50	0.52	0.46
136	7'' Light.	25	2.5	2.0	0.219	17.62	0.75	2.66	0.55	0.47
50	6'' Heavy.	56	5.6	2.69	0.625	26.50	3.02	2.18	0.73	0.79
50	6'' Light.	47	4.7	2.5	0.438	23.12	2.5	2.22	0.73	0.73
51	6'' Heavy.	36	3.6	2.20	0.383	16.89	1.31	2.17	0.60	0.56
51	6'' Light.	28	2.8	2.06	0.25	14.49	1.05	2.28	0.61	0.57
144	6'' Heavy.	28	2.8	1.86	0.281	12.39	0.67	2.10	0.49	0.41
144	6'' Light.	22	2.2	1.75	0.172	10.42	0.62	2.18	0.53	0.40
121	5'' Heavy.	30	3.0	2.06	0.375	10.17	0.93	1.84	0.56	0.56
121	5'' Light.	27	2.7	2.0	0.313	9.52	0.84	1.88	0.56	0.55
120	5'' Heavy.	21	2.1	1.83	0.268	7.19	0.52	1.85	0.50	0.45
120	5'' Light.	17	1.7	1.75	0.188	6.35	0.43	1.93	0.51	0.47
119	4'' Heavy.	27	2.7	2.06	0.375	5.87	0.90	1.47	0.58	0.59
119	4'' Light.	24	2.4	2.0	0.313	5.53	0.79	1.52	0.57	0.60
118	4'' Heavy.	18	1.8	1.83	0.263	4.14	0.48	1.52	0.52	0.50
118	4'' Light.	15	1.5	1.75	0.188	3.74	0.4	1.58	0.52	0.52
117	3'' Heavy.	18	1.8	1.63	0.375	2.26	0.36	1.12	0.45	0.53
116	3'' Light.	15	1.5	1.5	0.25	1.98	0.29	1.15	0.44	0.50

PHŒNIX ANGLE IRON.
EQUAL SIDES.

No of Shape.	DESIGNATION.	Weight Per Yard. Lbs.	Area of Section. Sq. In.	Thickness. Inches.	MOMENT OF INERTIA.		RADIUS OF GYRATION.		Distance of Centre of Gravity from Outside of Side.
					Neutral Axis through Centre of Gravity Parallel to Side.	Neutral Axis through Centre of Gravity at 45° to Sides.	Neutral Axis through Centre of Gravity Parallel to Side.	Neutral Axis through Centre of Gravity at 45° to Sides.	
127	6″ × 6″ Heavy.	93.5	9.35	0.813	29.62	12.15	1.78	1.14	1.70
127	6″ × 6″ Light.	50.3	5.03	0.438	17.22	6.77	1.85	1.16	1.58
126	5″ × 5″ Heavy.	62.	6.2	0.688	14.70	6.07	1.54	0.99	1.55
126	5″ × 5″ Light.	37.	3.7	0.406	9.35	3.77	1.59	1.01	1.46
14	4″ × 4″ Heavy.	51.6	5.16	0.688	7.18	3.01	1.18	0.76	1.22
14	4″ × 4″ Light.	28.1	2.81	0.375	4.39	1.71	1.25	0.78	1.16
15	3½″ × 3½″ Heavy.	41.	4.1	0.625	4.35	1.84	1.03	0.67	1.08
15	3½″ × 3½″ Light.	20.5	2.05	0.313	2.30	0.95	1.06	0.68	0.93
16	3″ × 3″ Heavy.	28.1	2.81	0.5	2.23	0.95	0.89	0.58	0.93

16	3'' × 3'' Light.	15.	1.5	0.25	1.33	0.54	0.94	0.6	0.87
37	2¾'' × 2¾'' Heavy.	25.8	2.58	0.5	1.65	0.62	0.80	0.49	0.83
37	2¾'' × 2¾'' Light.	13.4	1.34	0.25	1.01	0.41	0.87	0.55	0.82
17	2½'' × 2½'' Heavy.	23.6	2.36	0.5	1.22	0.52	0.72	0.47	0.77
17	2½'' × 2½'' Light.	10.5	1.05	0.219	0.62	0.25	0.77	0.49	0.7
38	2¼'' × 2¼'' Heavy.	18.3	1.83	0.438	0.82	0.35	0.67	0.44	0.74
38	2¼'' × 2¼'' Light.	8.	0.8	0.188	0.40	0.17	0.71	0.46	0.69
18	2'' × 2'' Heavy.	14.	1.4	0.375	0.49	0.20	0.59	0.38	0.62
18	2'' × 2'' Light.	7.5	0.75	0.188	0.29	0.12	0.62	0.40	0.6
19	1¾'' × 1¾'' Heavy.	10.1	1.01	0.313	0.27	0.12	0.52	0.34	0.55
19	1¾'' × 1¾'' Light.	6.1	0.61	0.188	0.18	0.07	0.55	0.35	0.52
20	1½'' × 1½'' Heavy.	7.1	0.71	0.25	0.14	0.06	0.44	0.29	0.45
20	1½'' × 1½'' Light.	4.4	0.44	0.156	0.09	0.04	0.46	0.29	0.44
39	1¼'' × 1¼'' Heavy.	4.3	0.43	0.188	0.06	0.02	0.38	0.24	0.36
39	1¼'' × 1¼'' Light.	2.8	0.28	0.125	0.05	0.02	0.42	0.28	0.43
40	1'' × 1'' Heavy.	3.6	0.36	0.188	0.03	0.01	0.29	0.19	0.31
40	1'' × 1'' Light.	2.4	0.24	0.125	0.02	0.01	0.30	0.19	0.29

PHŒNIX ANGLE IRON.
UNEQUAL SIDES.

No. of Shape.	DESIGNATION.	Weight Per Yard Lbs.	Area of Section. Sq. In.	Thickness. Inches.	MOMENT OF INERTIA.			RADIUS OF GYRATION.			DISTANCE OF CENTRE OF GRAVITY.	
					Neutral Axis Parallel to Long Side.	Neutral Axis Parallel to Short Side.	Neutral Axis Parallel to Line through Ends of Sides.	Neutral Axis Parallel to Long Side.	Neutral Axis Parallel to Short Side.	Neutral Axis Parallel to Line through Extremities of Sides.	From Long Side.	From Short Side.
87	6½" X 4" Heavy.	74.8	7.48	0.75	8.88	31.74	6.75	1.09	2.06	0.95	1.03	2.23
87	6½" X 4" Light.	40.7	4.07	0.406	5.38	17.61	3.60	1.15	2.08	0.94	0.91	2.18
91	6" X 4" Heavy.	71.2	7.12	0.75	8.93	24.63	5.77	1.12	1.86	0.90	1.08	2.10
91	6" X 4" Light.	36.5	3.65	0.375	5.00	13.60	2.89	1.17	1.93	0.89	0.97	2.00
92	6" X 3½" Heavy.	56.2	5.62	0.625	5.18	20.08	3.78	0.96	1.89	0.82	0.90	2.17
92	6" X 3½" Light.	33.8	3.38	0.375	3.38	12.59	2.11	1.00	1.93	0.79	0.82	2.11
41	5" X 4" Heavy.	53.1	5.31	0.625	7.27	12.76	3.93	1.17	1.55	0.86	1.11	1.60
41	5" X 4" Light.	31.9	3.19	0.375	4.59	8.06	2.20	1.20	1.59	0.83	1.04	1.55

93	5″ × 3½″ Heavy.	55.	5.5	0.688	5.28	13.38	3.52	0.98	1.56	0.80	0.94	1.63
93	5″ × 3½″ Light.	27.5	2.75	0.313	2.92	7.13	1.94	1.03	1.61	0.84	0.79	1.5
42	5″ × 3″ Heavy.	47.1	4.71	0.625	3.09	11.46	2.11	0.81	1.56	0.67	0.81	1.83
42	5″ × 3″ Light.	23.6	2.36	0.313	1.75	6.04	0.97	0.85	1.60	0.64	0.74	1.80
43	4½″ × 3″ Heavy.	39.7	3.97	0.563	2.80	7.89	1.89	0.84	1.41	0.69	0.82	1.56
43	4½″ × 3″ Light.	26.5	2.65	0.375	1.96	4.60	1.23	0.86	1.44	0.68	0.74	1.49
94	4″ × 3½″ Heavy.	39.7	3.97	0.563	4.21	5.89	2.29	1.03	1.22	0.76	0.99	1.23
94	4″ × 3½″ Light.	26.5	2.65	0.375	2.98	4.14	1.61	1.05	1.25	0.78	0.89	1.12
44	4″ × 3″ Heavy.	36.9	3.69	0.563	2.73	5.67	1.56	0.86	1.24	0.65	0.86	1.36
44	4″ × 3″ Light.	20.5	2.05	0.313	1.62	3.31	0.76	0.89	1.27	0.61	0.80	1.32
95	3½″ × 3″ Heavy.	34.1	3.41	0.563	2.58	3.90	1.35	0.87	1.07	0.63	0.88	1.12
95	3½″ × 3″ Light.	19.7	1.97	0.313	1.60	2.38	0.76	0.90	1.10	0.62	0.82	1.08
86	3″ × 2½″ Heavy.	25.8	2.58	0.5	1.34	2.28	0.81	0.72	0.94	0.56	0.72	0.95
86	3″ × 2½″ Light.	13.	1.3	0.25	0.72	1.25	0.36	0.75	0.98	0.52	0.68	0.92
109	3″ × 2″ Heavy.	17.8	1.78	0.375	0.56	1.54	0.39	0.56	0.93	0.47	0.53	1.02
109	3″ × 2″ Light.	11.9	1.19	0.25	0.38	1.09	0.25	0.57	0.96	0.46	0.49	0.99
96	2¼″ × 1¾″ Heavy.	9.	0.9	0.25	0.16	0.45	0.14	0.42	0.71	0.39	0.38	0.75
96	2¼″ × 1½″ Light.	7.5	0.75	0.188	0.13	0.49	0.12	0.42	0.81	0.40	0.33	0.67

PROPERTIES OF PHŒNIX TEE BARS.
UNEQUAL SIDES.

No. of Shape.	Size Flange by Web.	Weight Per Yard. Lbs.	Area. Sq. In.	Thickness of Web. Inches.	Thickness of Flange. Inches.	MOMENT OF INERTIA.		RADIUS OF GYRATION.		Distance of Centre of Gravity from Top.
						Neutral Axis Parallel to Flange.	Neutral Axis Coincident with Web.	Neutral Axis Parallel to Flange.	Neutral Axis Coincident with Web.	
23	5″ X 2½″	35	3.5	0.563	0.5	2.21	5.24	0.79	1.22	0.77
25	5″ X 2½″	29	2.9	0.563	0.375	1.39	3.94	0.69	1.17	0.66
132	4½″ X 3″	25	2.5	0.375	0.313	1.94	2.39	0.88	0.98	0.76
46	4″ X 3½″	49	4.9	0.797	0.625	6.50	3.47	1.15	0.84	1.27
85	4″ X 3½″	16.5	1.65	0.381	0.313	0.60	1.68	0.60	1.01	0.57
45	3″ X 3½″	32	3.2	0.5	0.469	4.17	1.08	1.14	0.58	1.18
24	3″ X 3½″	30	3.0	0.438	0.438	3.14	1.01	1.02	0.58	0.98
98	2½″ X 2½″	18	1.8	0.359	0.375	1.26	0.50	0.84	0.53	0.84
47	2⅝″ X 1⁷⁄₁₆″	6.5	0.65	0.297	0.188	0.86	0.15	0.36	0.15	0.37

EQUAL SIDES.

No. of Shape.	Size Flange by Web.	Weight Per Yard. Lbs.	Area. Sq. In.	Thickness of Web. Inches.	Thickness of Flange. Inches.	Neutral Axis Parallel to Flange.	Neutral Axis Coincident with Web.	Neutral Axis Parallel to Flange.	Neutral Axis Coincident with Web.	Distance of Centre of Gravity from Top.
101	3½″ X 3½″	28.5	2.85	0.422	0.453	3.20	1.64	1.06	0.76	1.02
102	3″ X 3″	21	2.1	0.375	0.375	1.76	0.86	0.92	0.69	0.89
84	2½″ X 2½″	16	1.6	0.344	0.344	0.92	0.46	0.76	0.53	0.75
103	2″ X 2″	9	0.9	0.25	0.25	0.35	0.17	0.20	0.14	0.62

DETAILS OF CONSTRUCTION

IN

WROUGHT-IRON WORK.

FOR the convenience of Architects, Engineers, and Builders, some of the details of construction employed in wrought-iron work are given in the following pages, and the adaptations of the various shapes to structural uses will be illustrated and explained under the several heads into which the work is classified.

In the building of FLOORS and ROOFS, it is customary to make use of BEAMS, CHANNELS, COLUMNS, and other shapes of rolled iron.

FLOORS.

In planning a floor, the first point to be determined is the load that will probably be placed upon it.

The weight of the materials composing the floor is usually termed the *dead* load, and the weight of the persons or stores of any kind that may be placed upon the floor is called the *live* load. The dead load of a fire-proof floor, made of rolled beams and four-inch brick arches, filled in above with concrete, may be taken at 70 pounds per square foot, and the live load for dwellings or offices may be assumed at 70 pounds additional, and on these assumptions the table on page 85 has been calculated. But in public buildings or churches, where large crowds of persons in motion may congregate, or in warehouses where heavy goods may be stored, it is evident that the loads will have to be determined by the circumstances, and will exceed the amounts above specified.

For ordinary conditions the following total loads per square foot may be assumed as giving a safe approximation in practice:

Dwellings or Office Buildings . .	140 pounds.
Public Halls or Churches	175 "
Warehouses	150 to 300 "

In order to support these loads with entire safety, I beams of various dimensions are offered in the accompanying tables. For floors of small span the lighter beams can be economically used, but for greater spans larger beams are necessary.

That a beam should be strong enough to support a given load for a given span is not all that is requisite—it is equally important that it should be stiff enough. Rigidity prevents vibration, and the avoidance of this is of great importance, since repeated movements in the floor would injure and possibly destroy the masonry in the brick-work. It is, therefore, advisable, where circumstances permit, to consider whether deep beams placed further apart might not prove to be more economical than light beams near to each other.

For the proper spacing of beams under various loads, reference may be had to the diagram given on page 40.

Under no circumstances, however, should beams be strained beyond the limits of their elasticity; or, in other words, so strained that on the removal of the load they will not return to their original condition without set or permanent deflexion.

If a beam is required to sustain a load concentrated at the centre of the span, it must be noted that only one-half as much weight can be borne when so concentrated as could be supported if the load were uniformly distributed over the whole beam.

The figures given in the tables for the load-bearing capacity of any beam must then be divided by 2 to ascertain the safe load concentrated at the middle of the span, and this concentrated load will cause the beam to deflect $\frac{8}{10}$ as much as would the distributed load named.

If the load is to be concentrated at any other point than the centre, then the following statement of proportion will determine the case: The weight that the beam can carry at the centre is to the weight that it can carry at any other point as the rectangle of the segments of the span at the given point is to the square of half the span. For example, supposing a 12-inch 125-pound beam to support with safety a central load of five tons for a span of 20 feet, what load will it carry concentrated at a point 5 feet from one wall?

Here, 5 tons : X tons :: 5 $\times$ 15 : 10 $\times$ 10, or $6\frac{2}{3}$ tons.

This rule is of service in such cases as when it is required to provide proper beams in floors under heavy local loads, such as safes or vaults.

Having determined the load per square foot to be sustained, the proper beams to use may be ascertained by reference to Table II. The coefficient of safety is placed above each beam in this table, and this divided by the clear span in feet will show the strength of the beam at this span for a distributed load in net tons of 2000 pounds. The deflexion of the beam corresponding to this load will be found in the next line, and the weight of the beam should be deducted from the safe load. For any less load uniformly distributed the deflexion will be directly proportionate to that given in the table.

To determine the strength of beams many experiments have been made, and the generally accepted theory with regard to the effect of applied loads is that which assumes a neutral axis at the centre of gravity of the cross-section of the beam, and supposes the material above this axis to be compressed while that below the axis is extended, the resistance of any element to the strains of compression or extension being directly as its distance from the neutral axis.

Certain general principles have been fully confirmed by experiment, such, for instance, as that in beams of equal length and breadth the strength varies directly as the square of the depth, and in beams of equal length and depth directly as the breadth.

Hence the strength of any beam may be represented by the following expression:

$$W = \frac{breadth \times square\ of\ depth}{length} \times constant.$$

The value of the constant being dependent upon the material of the beam. This may also be written,

$$W = \frac{area \times depth \times constant}{length} = \frac{a \times d \times c}{L}.$$

Representing the various conditions of loading, it has further been determined by experiment that the following proportions obtain for all beams

Fixed at one end and loaded at the other,

$$W = \frac{a \times d \times c}{L};$$

Fixed at one end and uniformly loaded,

$$W = 2 \left(\frac{a \times d \times c}{L} \right).$$

Supported at both ends and centrally loaded,

$$W = 4 \left(\frac{a \times d \times c}{L} \right);$$

Supported at both ends and uniformly loaded,

$$W = 8 \left(\frac{a \times d \times c}{L} \right).$$

To apply these formulæ to any given beam, it is necessary to obtain by experiment the value of the constant c, taking the average of a number of tests. One-sixth, one-fourth, or even one-third of this value may be taken as the working load, according to the conditions of service for which the beam may be designed. For wrought-iron rolled beams, c may be taken as 48,000 pounds, and the safe load per square inch of effective section at 12,000 pounds, or six net tons, and with this as a constant the tables showing the strength of Phœnix beams have been computed.

By "effective section" is meant that portion of the total section which is effective in resisting the strains of tension or compression, and it is ordinarily computed by adding one-sixth of the area of the stem or web to the entire area of one flange; thus, $a + \frac{a'}{6}$.

In this estimate of the effective section two-thirds of the area of the web have been omitted from the calculation, because of the assumption that this portion of the web lies too near to the neutral axis to assist in offering any resistance to the strains caused by a load.

The "effective depth" of a beam is the distance between the centres of gravity of its two flanges, and in Table I this effective depth has been expressed, both in feet, D, and in inches, d; the former being required in the formula for strength, while the latter is required in the formula for deflexion.

For rolled beams, under the equally distributed loads of floors, the effective section of the lower flange is in tension and the upper flange in compression, so that if the safe load of six tons per square inch is assumed, the general formula will be

$$W = 8 \left(\frac{a \times d \times c}{L} \right) = \frac{8\,D\left(a + \frac{a'}{6}\right)6.}{L}$$

Now, in this formula, it is only necessary to insert the proper values for "effective depth" and "effective section" given in the table for each particular beam, in order to determine its strength for any given span. The load-factor for each beam is thus dependent upon its depth and the quantity of metal in its flanges. This load-factor, when divided by the number expressing the clear span in feet, will give as a quotient a number indicating the weight in tons that the beam will carry with safety. For the several beams, the tables show what the proper loads are that may be placed upon them for each foot of clear span.

Stiffness is a different quality from strength. A beam that may be quite strong enough to carry a given load may deflect under this load more than is desirable.

About one-thirtieth of an inch per foot of clear span is the usual maximum of deflexion that is permissible. Under ordinary loads this is attained when the clear span is about twenty-six times the depth of the beam, and the heavy lines in the tables show for each beam where this limit may be found.

Like the load-factor, the bending moment is dependent upon the effective depth and the effective section of the beam to which it is to be applied; the general formula for the deflexion of any beam under an equally distributed load

being $\delta' = \dfrac{.004 \; W. \; L^3}{\left(a + \frac{a'}{6}\right) d^2}$.

By inserting the values proper to each beam, the results given in the following tables have been obtained. For the process of deriving this formula, see page 76 following. A close approximation to the actual deflexion at the centre, under the maximum safe load, may be obtained by dividing the square of the length of the span in feet by 62 times the depth of the beam in inches.

DEFINITION OF TERMS USED IN FORMULÆ.

W = Equally distributed load on any beam in net tons.
L = Length of clear span, expressed in feet.
a = Area of top, or bottom, flange, in square inches.
a′ = Area of stem of beam, in square inches.
D = Effective depth of beam, expressed in feet.
d = Effective depth of beam, expressed in inches.
S = Strain per square inch of effective section $\left(a + \frac{a'}{6}\right)$ in tons of 2000 pounds.
δ = Deflexion in inches at middle for a central load.
δ′ = Deflexion in inches at middle for a uniformly distributed load.

General formula for any I beam under an equally distributed load. $\} \; W = \dfrac{8 \; D \; \left(a + \frac{a'}{6}\right) S}{L}$

TABLE I.

ELEMENTS OF PHŒNIX BEAMS.

BEAM	DIMENSIONS, INCHES			AREA, SQUARE INCHES			EFFECTIVE DEPTH		Load Factor	Deflection Factor
	Width of Flange	Average Thickness of Flange	Thickness of Stem	a of Flange	a' of Stem	Sum of $a+\frac{a'}{6}$	D Feet	d Inches	$8D\left(a+\frac{a'}{6}\right)S$ When $S'=6$ Tons	$\left(a+\frac{a'}{6}\right)d^2$
15″ 200	5 5/16	1.156	.65	6.142	7.715	7.428	1.150	13.80	410	1415
15″ 150	4 5/8	.911	.50	4.330	6.340	5.386	1.170	14.04	302	1062
15″ 125	4 3/8	.734	.42	3.395	5.710	4.347	1.188	14.26	248	884
12″ 170	5 1/4	1.050	.59	5.777	5.446	6.684	.910	10.92	292	797
12″ 125	4 5/8	.802	.49	3.810	4.880	4.623	.930	11.16	208	576
12″ 96	4 1/2	.609	.38	2.740	4.119	3.426	.949	11.39	156	444
10½″ 135	5	.875	.50	4.375	4.750	5.166	.800	9.62	178	478
10½″ 105	4 3/4	.745	.44	3.353	3.793	3.986	.812	9.74	155	378
10½″ 90	4 1/2	.640	.37	2.800	3.400	3.366	.825	9.87	133	327
9″ 150	5 5/8	1.039	.60	5.586	3.828	6.224	.658	7.90	197	388
9″ 84	4	.700	.40	2.800	2.800	3.261	.691	8.30	108	225
9″ 70	3 1/2	.680	.31	2.381	2.238	2.754	.698	8.38	92	193
8″ 81	4 1/2	.625	.38	2.812	2.476	3.225	.610	7.37	94	175
8″ 65	4	.527	.35	2.109	2.282	2.489	.618	7.42	74	137
7″ 69	4	.625	.37	2.500	1.900	2.816	.530	6.37	72	114
7″ 55	3 1/2	.507	.35	1.775	1.949	2.100	.537	6.44	54	87
6″ 50	3 1/2	.531	.31	1.858	1.284	2.072	.456	5.47	45	62
6″ 40	3 1/4	.517	.25	1.421	1.158	1.614	.458	5.50	35	49
5″ 40	3	.400	.30	1.200	1.200	1.400	.383	4.60	25	30
5″ 30	2 3/4	.375	.25	1.000	1.000	1.166	.385	4.62	21	25
4″ 30	2 3/4	.410	.25	1.135	.730	1.257	.298	3.58	18	16
4″ 18	2	.281	.21	.562	.682	.676	.304	3.65	10	9

The general formulæ for deflexions given below are taken from Professor Moseley's "Mechanics of Engineering," edited by Professor Mahan, in 1856, changing the letters which he has employed to agree with those used in this work.

Let l = The clear span, in inches.

E = Modulus of elasticity = 24,000,000 pounds = 12,000 tons.

I = Moment of inertia for the several forms.

δ = Deflexion at middle, in inches.

W = Load, in tons, producing deflexion.

a = Area, and d = depth of beam, in inches.

Then, for a beam fixed at one end and loaded at the other,

$$\delta = \frac{W\,l^3}{3\,E\,I}$$

For a beam fixed at one end and uniformly loaded,

$$\delta = \frac{W\,l^3}{8\,E\,I}$$

For a beam supported at both ends and loaded at the centre,

$$\delta = \frac{W\,l^3}{48\,E\,I}$$

For a beam supported at both ends and uniformly loaded,

$$\delta = \tfrac{5}{8} \times \frac{W\,l^3}{48\,E\,I}$$

For the several sections of beams the value of I will be as follows :

1. $\quad I = \dfrac{b\,d^3}{12}$

4. $\quad I = .7854\,(r^4 - r'^4)$

2. $\quad I = \dfrac{b\,d^3 - b'\,d'^3}{12}$

5. $\quad I = \tfrac{1}{6}\left\{ bd^3 + b'd'^3 - (b'-b)d''^3 \right\}$

3. $\quad I = .7854\,r^4 - \dfrac{a\,r^2}{4}$

6. $\quad I = \dfrac{b\,d^3 - b'\,d'^3}{12}$

By substituting, in formula 6, the effective areas of flange and stem,

$$I = \frac{d^2}{12}\,(6\,a + a')$$

Then, for shape 6, supported at both ends and loaded at the centre,

$$\delta = \frac{W\,l^3}{48 \times 12,000 \times \dfrac{d^2}{12}\,(6\,a + a')}$$

Substituting 1728 L³ for l^3, to express the length of span in feet instead of inches, we have :

$$\delta = \frac{W\,L^3}{27.78\,(6\,a + a')\,d^2} = \frac{.036\,W\,L^3}{(6\,a + a')\,d^2} = \frac{.006\,W\,L^3}{\left(a + \dfrac{a'}{6}\right)d^2}$$

And for shape 6, supported at both ends and uniformly loaded,

$$\delta = \frac{.004\,W\,L^2}{\left(a + \dfrac{a'}{6}\right)d^2}$$

In this form the formula for deflexion will be found in the table of beams, Table I.

TABLES OF BEAMS,

SHOWING THE PROPER SIZES FOR

Varying Conditions of Loading and Spacing,

WITH THE CORRESPONDING

DEFLEXIONS UNDER THE SAFE LOADS.

TABLE II.

COMPARATIVE STRENGTH AND STIFFNESS

OF THE

DIFFERENT SECTIONS OF WROUGHT-IRON BEAMS,

MADE BY THE

PHŒNIX IRON COMPANY,

FOR

Sustaining, with entire safety, a Uniformly Distributed Load.

Clear Span, in Feet.	1 — 15″ — 200 Lbs. — $W = \frac{410}{L}$			89 — 15″ — 150 Lbs. — $W = \frac{302}{L}$			138 — 15″ — 125 Lbs. — $W = \frac{248}{L}$		
	Safe Load, Net Tons.	Correspon'g Deflexion.	Wt. of Beam, in Lbs.	Safe Load, Net Tons.	Correspon'g Deflexion.	Wt. of Beam, in Lbs.	Safe Load, Net Tons.	Correspon'g Deflexion.	Wt. of Beam, in Lbs.
10	41.0	.116	667	30.2	.114	500	24.8	.112	417
11	37.2	.140	733	27.4	.138	550	22.5	.135	458
12	34.2	.167	800	25.2	.154	600	20.7	.162	500
13	31.6	.196	867	23.2	.182	650	19.0	.189	542
14	29.3	.227	933	21.6	.212	700	17.7	.219	583
15	27.4	.261	1000	20.0	.254	750	16.6	.253	625
16	25.6	.296	1067	18.9	.289	800	15.5	.287	667
17	24.1	.334	1133	17.8	.327	850	14.6	.324	708
18	22.8	.376	1200	16.8	.367	900	13.8	.364	750
19	21.6	.419	1267	15.9	.410	950	13.0	.403	792
20	20.5	.463	1333	15.1	.455	1000	12.4	.449	833
21	19.5	.510	1400	14.4	.502	1050	11.8	.494	875
22	18.6	.560	1467	13.7	.551	1100	11.2	.539	917
23	17.8	.612	1533	13.1	.602	1150	10.7	.589	958
24	17.1	.667	1600	12.6	.656	1200	10.3	.644	1000
25	16.4	.725	1667	12.1	.712	1250	9.9	.699	1042
26	15.8	.785	1733	11.6	.769	1300	9.5	.755	1083
27	15.2	.846	1800	11.2	.828	1350	9.2	.819	1125
28	14.6	.906	1867	10.8	.889	1400	8.9	.884	1167
29	14.1	.972	1933	10.4	.942	1450	8.6	.966	1208
30	13.7	1.040	2000	10.0	1.017	1500	8.3	1.014	1250

TABLE II.

COMPARATIVE STRENGTH AND STIFFNESS

OF THE

DIFFERENT SECTIONS OF WROUGHT-IRON BEAMS,

MADE BY THE

PHŒNIX IRON COMPANY,

FOR

Sustaining, with entire safety, a Uniformly Distributed Load.

55 — 12″ 170 Lbs. $W=\frac{292}{L}$			57 — 12″ 125 Lbs. $W=\frac{208}{L}$			139 — 12″ 96 Lbs. $W=\frac{156}{L}$			
Safe Load, Net Tons.	Cor:espon'g Deflection.	Wt. of Beam, in Lbs.	Safe Load, Net Tons.	Cor+espon'g Deflection.	Wt. of Beam, in Lbs.	Safe Load, Net Tons.	Corcespon'g Deflection.	Wt. of Beam, in Lbs.	Clear Span, in Feet.
29.2	.147	567	20.8	.144	417	15.6	.140	320	10
26.6	.177	623	18.8	.174	458	14.2	.170	352	11
24.3	.210	680	17.3	.207	500	13.0	.202	384	12
22.4	.246	737	16.0	.243	542	12.0	.237	416	13
20.9	.286	793	14.9	.282	583	11.1	.252	448	14
19.4	.328	850	13.8	.325	625	10.4	.316	480	15
18.3	.374	907	13.0	.360	667	9.7	.351	512	16
17.2	.423	963	12.2	.408	708	9.2	.407	544	17
16.2	.475	1020	11.5	.459	750	8.7	.457	576	18
15.4	.530	1077	10.9	.513	792	8.2	.537	608	19
14.6	.587	1133	10.4	.578	833	7.8	.562	640	20
13.9	.648	1190	9.9	.636	875	7.4	.617	672	21
13.3	.711	1247	9.4	.698	917	7.1	.685	704	22
12.7	.777	1303	9.0	.763	958	6.8	.744	736	23
12.2	.846	1360	8.7	.832	1000	6.5	.809	768	24
11.7	.918	1417	8.3	.903	1042	6.2	.872	800	25
11.2	.992	1473	8.0	.997	1083	6.0	.950	832	26
10.8	1.068	1530	7.7	1.053	1125	5.7	1.010	864	27
10.4	1.147	1587	7.4	1.131	1167	5.5	1.087	896	28
10.0	1.230	1643	7.1	1.211	1208	5.3	1.186	928	29
9.7	1.314	1700	6.9	1.294	1250	5.2	1.265	960	30

TABLE II.

COMPARATIVE STRENGTH AND STIFFNESS

OF THE

DIFFERENT SECTIONS OF WROUGHT-IRON BEAMS,

MADE BY THE

PHŒNIX IRON COMPANY,

FOR

Sustaining, with entire safety, a Uniformly Distributed Load.

Clear Span, in Feet.	114 10½" 135 Lbs. $W = \dfrac{178}{L}$			58 10½" 105 Lbs. $W = \dfrac{155}{L}$			131 10½" 90 Lbs. $W = \dfrac{133}{L}$		
	Safe Load, Net Tons.	Correspon'g Deflexion.	Wt. of Beam, in Lbs.	Safe Load, Net Tons.	Correspon'g Deflexion.	Wt. of Beam, in Lbs.	Safe Load, Net Tons.	Correspon'g Deflexion.	Wt. of Beam, in Lbs.
		″			″			″	
10	17.8	.149	450	15.5	.164	350	13.3	.162	300
11	16.2	.180	495	14.0	.197	385	12.1	.197	330
12	14.8	.214	540	12.9	.236	420	11.0	.232	360
13	13.7	.251	585	11.8	.278	455	10.2	.274	390
14	12.7	.291	630	11.1	.322	490	9.5	.318	420
15	11.8	.333	675	10.2	.364	525	8.8	.363	450
16	11.1	.380	720	9.7	.414	560	8.3	.415	480
17	10.5	.431	765	9.1	.470	595	7.8	.468	510
18	9.9	.481	810	8.6	.528	630	7.4	.527	540
19	9.3	.533	855	8.1	.589	665	7.0	.587	570
20	8.9	.595	900	7.7	.652	700	6.6	.645	600
21	8.5	.658	945	7.3	.719	735	6.3	.713	630
22	8.1	.721	990	7.0	.788	770	6.0	.781	660
23	7.7	.784	1035	6.7	.862	805	5.7	.848	690
24	7.4	.856	1080	6.5	.941	840	5.5	.930	720
25	7.1	.928	1125	6.2	1.025	875	5.3	1.013	750
26	6.8	1.00	1170	5.9	1.105	910	5.1	1.096	780
27	6.6	1.08	1215	5.7	1.187	945	4.9	1.179	810
28	6.3	1.16	1260	5.5	1.271	980	4.7	1.262	840
29	6.1	1.24	1305	5.3	1.360	1015	4.6	1.372	870
30	5.9	1.33	1350	5.1	1.455	1050	4.4	1.453	900

TABLE II.

COMPARATIVE STRENGTH AND STIFFNESS

OF THE

DIFFERENT SECTIONS OF WROUGHT-IRON BEAMS,

MADE BY THE

PHŒNIX IRON COMPANY,

FOR

Sustaining, with entire safety, a Uniformly Distributed Load.

	4 9″ 150 Lbs. $W = \frac{197}{L}$			5 9″ 84 Lbs. $W = \frac{108}{L}$			6 9″ 70 Lbs. $W = \frac{92}{L}$		
Clear Span, in Feet.	Safe Load, Net Tons.	Corresponding Deflection.	Wt. of Beam, in Lbs.	Safe Load, Net Tons.	Corresponding Deflection.	Wt. of Beam, in Lbs.	Safe Load, Net Tons.	Corresponding Deflection.	Wt. of Beam, in Lbs.
		″			″			″	
10	19.7	.203	500	10.8	.192	280	9.2	.190	233
11	17.8	.243	550	9.8	.231	308	8.4	.231	256
12	16.4	.296	600	9.0	.276	336	7.7	.275	280
13	15.2	.347	650	8.3	.324	364	7.0	.318	303
14	14.1	.402	700	7.7	.376	392	6.7	.380	326
15	13.2	.459	750	7.2	.432	420	6.2	.432	350
16	12.3	.530	800	6.7	.488	448	5.7	.448	373
17	11.6	.585	850	6.3	.550	476	5.4	.548	396
18	10.9	.654	900	6.0	.622	504	5.1	.615	420
19	10.3	.737	950	5.7	.695	532	4.8	.690	443
20	9.8	.807	1000	5.4	.768	560	4.6	.761	466
21	9.3	.891	1050	5.1	.839	588	4.4	.842	490
22	8.9	.980	1100	4.9	.927	616	4.2	.925	513
23	8.5	1.07	1150	4.7	1.01	644	4.0	1.01	536
24	8.2	1.17	1200	4.5	1.10	672	3.8	1.08	560
25	7.9	1.27	1250	4.3	1.19	700	3.6	1.16	583
26	7.6	1.38	1300	4.1	1.27	728	3.5	1.27	606
27	7.3	1.48	1350	3.9	1.36	756	3.4	1.38	630
28	7.0	1.59	1400	3.8	1.48	784	3.3	1.49	653
29	6.8	1.70	1450	3.7	1.60	812	3.2	1.60	676
30	6.6	1.83	1500	3.6	1.73	840	3.1	1.73	700

TABLE II.
COMPARATIVE STRENGTH AND STIFFNESS
OF THE
DIFFERENT SECTIONS OF WROUGHT-IRON BEAMS,
MADE BY THE
PHŒNIX IRON COMPANY,
FOR
Sustaining, with entire safety, a Uniformly Distributed Load.

113 8″ 81 Lbs. $W = \frac{94}{L}$			59 8″ 65 Lbs. $W = \frac{74}{L}$			112 7″ 69 Lbs. $W = \frac{72}{L}$			
Safe Load, Net Tons.	Correspon'g Deflexion.	Wt. of Beam, in Lbs.	Safe Load, Net Tons.	Correspon'g Deflexion.	Wt. of Beam, in Lbs.	Safe Load, Net Tons.	Correspon'g Deflexion.	Wt. of Beam, in Lbs.	Clear Span, in Feet.
9.4	.215	270	7.4	.215	216	7.2	.252	230	10
8.5	.258	297	6.8	.264	238	6.5	.303	253	11
7.8	.308	324	6.2	.312	260	6.0	.363	276	12
7.2	.361	351	5.7	.365	282	5.5	.424	299	13
6.7	.420	378	5.3	.424	303	5.1	.491	322	14
6.2	.478	405	4.9	.475	325	4.8	.568	345	15
5.9	.546	432	4.6	.549	347	4.5	.645	368	16
5.5	.617	459	4.3	.616	368	4.2	.724	391	17
5.2	.693	486	4.1	.697	390	4.0	.818	414	18
5.0	.783	513	3.9	.780	412	3.8	.914	437	19
4.7	.859	540	3.7	.863	433	3.6	1.01	460	20
4.5	.952	567	3.5	.946	455	3.4	1.10	483	21
4.2	1.02	594	3.4	1.05	477	3.2	1.19	506	22
4.1	1.14	621	3.2	1.13	498	3.1	1.32	529	23
3.9	1.23	648	3.1	1.25	520	3.0	1.45	552	24
3.7	1.32	675	2.9	1.32	542	2.9	1.59	575	25
3.6	1.44	702	2.8	1.43	563	2.8	1.72	598	26
3.5	1.57	729	2.7	1.55	585	2.7	1.86	621	27
3.3	1.65	756	2.6	1.66	607	2.6	2.00	644	28
3.2	1.78	783	2.5	1.77	628	2.5	2.14	667	29
3.1	1.91	810	2.4	1.88	650	2.4	2.27	690	30

TABLE II.

COMPARATIVE STRENGTH AND STIFFNESS

OF THE

DIFFERENT SECTIONS OF WROUGHT-IRON BEAMS,

MADE BY THE

PHŒNIX IRON COMPANY,

FOR

Sustaining, with entire safety, a Uniformly Distributed Load.

Clear Span, in Feet	7" 55 Lbs. $W = \frac{54}{L}$			6" 50 Lbs. $W = \frac{45}{L}$			6" 40 Lbs. $W = \frac{35}{L}$		
	Safe Load, Net Tons.	Correspon'g Deflection.	Wt. of Beam, in Lbs.	Safe Load, Net Tons.	Correspon'g Deflection.	Wt. of Beam, in Lbs.	Safe Load, Net Tons.	Correspon'g Deflection.	Wt. of Beam, in Lbs.
10	5.4	.248	183	4.5	.290	167	3.5	.286	133
11	4.8	.293	201	4.1	.352	183	3.2	.348	146
12	4.5	.357	220	3.7	.412	200	2.9	.410	160
13	4.2	.423	238	3.4	.481	217	2.7	.486	173
14	3.9	.491	256	3.2	.566	233	2.5	.562	186
15	3.6	.558	275	3.0	.653	250	2.3	.636	200
16	3.4	.651	293	2.8	.740	267	2.2	.738	213
17	3.2	.722	311	2.6	.824	283	2.0	.805	226
18	3.0	.803	330	2.5	.940	300	1.9	.907	240
19	2.8	.882	348	2.4	1.06	317	1.8	1.01	253
20	2.7	.992	366	2.2	1.13	333	1.7	1.11	266
21	2.5	1.06	385	2.1	1.25	350	1.6	1.21	280
22	2.4	1.17	403	2.0	1.37	367	1.6	1.39	293
23	2.3	1.28	421	1.9	1.49	383	1.5	1.49	306
24	2.2	1.39	440	1.8	1.60	400	1.5	1.58	320
25	2.1	1.50	458	1.8	1.81	417	1.4	1.79	333
26	2.1	1.69	476	1.7	1.92	433	1.3	1.87	346
27	2.0	1.80	495	1.6	2.02	450	1.3	2.09	360
28	1.9	1.90	513	1.6	2.26	467	1.2	2.15	373
29	1.8	2.01	531	1.5	2.36	483	1.2	2.39	386
30	1.8	2.23	550	1.5	2.61	500	1.1	2.43	400

TABLE II.

COMPARATIVE STRENGTH AND STIFFNESS

OF THE

DIFFERENT SECTIONS OF WROUGHT-IRON BEAMS,

MADE BY THE

PHŒNIX IRON COMPANY,

FOR

Sustaining, with entire safety, a Uniformly Distributed Load.

106 5″ 36 Lbs. $W = \frac{75}{L}$			105 5″ 30 Lbs. $W = \frac{21}{L}$			65 4″ 30 Lbs. $W = \frac{18}{L}$			
Safe Load, Net Tons.	Correspon'g Deflexion.	Wt. of Beam, in Lbs.	Safe Load, Net Tons.	Correspon'g Deflexion.	Wt. of Beam, in Lbs.	Safe Load, Net Tons.	Correspon'g Deflexion.	Wt. of Beam, in Lbs.	Clear Span, in Feet.
2.5	.337	120	2.1	.336	100	1.80	.448	100	10
2.3	.413	132	1.9	.405	110	1.63	.545	110	11
2.0	.466	144	1.7	.471	120	1.50	.643	120	12
1.9	.563	156	1.6	.563	130	1.38	.752	130	13
1.8	.667	168	1.5	.660	140	1.28	.872	140	14
1.7	.774	180	1.4	.757	150	1.20	1.00	150	15
1.6	.885	192	1.3	.854	160	1.12	1.13	160	16
1.5	.995	204	1.2	.945	170	1.06	1.29	170	17
1.4	1.10	216	1.2	1.12	180	1.00	1.44	180	18
1.3	1.20	228	1.1	1.21	190	.95	1.62	190	19
1.2	1.29	240	1.0	1.28	200	.90	1.79	200	20
1.2	1.50	252	1.0	1.45	210	.85	1.95	210	21
1.1	1.58	264	.95	1.62	220	.81	2.14	220	22
1.1	1.80	276	.90	1.75	230	.78	2.35	230	23
1.0	1.86	288	.85	1.88	240	.75	2.57	240	24
1.0	2.11	300	.82	2.05	250	.72	2.79	250	25
.95	2.25	312	.80	2.25	260	.69	3.01	260	26
.92	2.44	324	.77	2.43	270	.66	3.26	270	27
.90	2.66	336	.75	2.64	280	.64	3.51	280	28
.86	2.83	348	.72	2.81	290	.62	3.77	290	29
.83	3.20	360	.70	3.03	300	.60	4.02	300	30

PHŒNIX BEAMS.

THEIR ADAPTATION AND DUTY AS FLOORING JOISTS.

Clear Span.	3' apart	3½' apart	4' apart	4½' apart	5' apart	5½' apart	6' apart
10 feet.	30 □'	35 □'	40 □'	45 □'	50 □'	55 □'	60 □'
Load lbs.	4,200	4,900	5,600	6,300	7,000	7,700	8,400
I		6"				7 or 8"	
12 feet.	36 □'	42	48	54	60	66	72
Load lbs.	5,040	5,880	6,720	7,560	8,400	9,240	10,080
I	6 or 7"			7"		8"	
14 feet.	42 □'	49	56	63	70	77	84
Load lbs.	5,880	6,860	7,840	8,820	9,800	10,780	11,760
I	7 or 8"		8 or 9" 70		9" 70		
16 feet.	48 □'	56	64	72	80	88	96
Load lbs.	6,720	7,840	8,960	10,080	11,200	12,320	13,440
I	8"		9" 70	9" 84		10½" 105	
18 feet.	54 □'	63	72	81	90	99	108
Load lbs.	7,560	8,820	10,080	11,340	12,600	13,860	15,120
I	8 or 9" 70	9" 84			10½" 105		
20 feet.	60 □'	70	80	90	100	110	120
Load lbs.	8,400	9,800	11,200	12,600	14,000	15,400	16,800
I	9 84 or 10½		10½" 105			12" 125	
22 feet.	66 □'	77	88	99	110	121	132
Load lbs.	9,240	10,780	12,320	13,860	15,400	16,940	18,480
I		10½" 105			12" 125		12" 170
24 feet.	72 □'	84	96	108	120	132	144
Load lbs.	10,080	11,760	13,440	15,120	16,800	18,480	20,160
I	10½ or 12" 125			12" 125		12" 170 or 15" 150	
26 feet.	78 □'	91	104	117	130	143	156
Load lbs.	10,928	12,740	14,560	16,380	18,240	20,020	21,840
I	10½ or 12		12" 125		12" 170 or 15" 150		15" 150
28 feet.	84' □	98	112	126	140	154	168
Load lbs.	11,760	13,720	15,680	17,640	19,600	21,560	23,520
I	12" 125 or 15" 150		12" 170 or 15" 150		15" 150		15" 200
30 feet.	90 □'	105	120	135	150	165	180
Load lbs.	12,600	14,700	16,800	18,900	21,000	23,100	25,200
I	12 or 15 150	12" 170 or 15" 150			15" 150		15" 200

In above table the load is taken at 140 lbs. per □ foot of floor.

STANDARD

BOLTS AND CAST SEPARATORS FOR COMPOUND BEAMS.

Number and Size of Beams.			C. to C. of Beams.	C. to C. of Bolts.	Weight in lbs.		Sizes of Bolts.		Length of Sepa'r.	O. to O. of Beam Flanges
					Cast Sepa'r.	Two Bolts.	Diam.	Length.		
2	15''	200	6''	9''	19	3 1/4	3/4''	7 5/8''	5 3/8''	11 1/4''
2	15	150	5 1/4	9	17	3	...	6 3/4	4 3/4	10
2	15	125	5	9	17	3	...	6 1/2	4 5/8	9 5/8
2	12	170	6	6 1/4	15	3 1/4	...	7 3/8	5 3/8	11 1/2
2	12	125	5 1/4	6 1/4	15	3	...	6 5/8	4 1/4	10
2	12	96	5	6 1/4	15	3	...	6 1/4	4 3/8	9 1/4
2	10 1/2	135	5 1/2	5 1/4	11	3	...	7	5	10 1/2
2	10 1/2	105	5	5 1/4	11	3	...	6 1/2	4 5/8	9 1/2
2	10 1/2	90	5	5 1/4	11	3	...	6 1/4	4 3/8	9 1/4
2	9	150	6	4 1/2	9	3 1/4	...	7 3/8	5 1/8	11 1/4
2	9	84	4 1/2	4 1/2	9	2 1/2	...	6	4 1/4	8 1/2
2	9	70	4	4 1/2	9	2 1/4	...	5 3/8	3 3/4	7 1/2
2	8	81	5	4	8	2 1/4	5/8''	6 1/4	4 3/8	9 1/4
2	8	65	4 1/4	4	8	2	...	5 1/4	4 1/4	8 1/2
2	7	69	4 1/2	3	7	1 3/4	...	5 1/4	4 1/8	8 1/4
2	7	55	4	3	7	1 3/4	...	5 1/4	3 5/8	7 1/2
2	6	50	4	3	5	1 1/2	...	5 1/4	3 1/4	7 1/4
2	6	40	3	3	5	1 1/2	...	4 5/8	2 3/4	5 3/4

STANDARD BRACKETS FOR BEAMS.

Size of Beam.	Brackets.		Bolts.		Rivets.		Approx Wt. of 1 Set.
	No.	Size of L	No.	Size.	No.	Size.	
15''	2	4 × 4 − 10''	6	3/4 × 2''	3	3/4 × 2 7/8	26
12	2	3 1/2 × 3 1/2 − 7 1/2	6	3/4 × 1 7/8	3	3/4 × 2 1/2	17
10 1/2	2	3 1/2 × 3 1/2 − 7 1/2	6	3/4 × 1 7/8	3	3/4 × 2 1/2	17
9	2	3 × 3 − 5 1/2	4	5/8 × 1 3/4	2	5/8 × 2 1/4	9
8	2	3 × 3 − 5 1/2	4	5/8 × 1 3/4	2	5/8 × 2 1/4	9
7	2	3 × 3 − 4	4	5/8 × 1 3/4	2	5/8 × 2 1/4	7 1/2
6	2	3 × 3 − 4	4	5/8 × 1 3/4	2	5/8 × 2 1/4	7 1/2

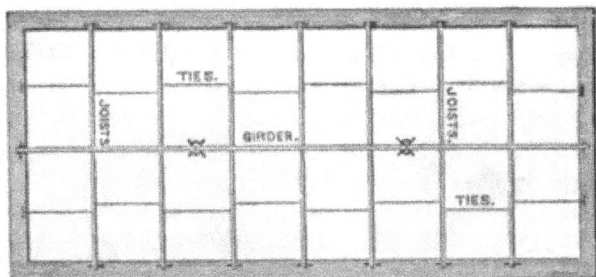

Fig.1

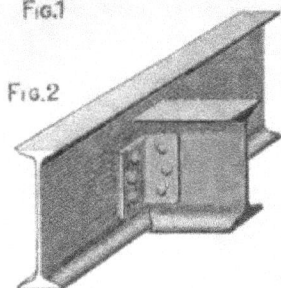

Fig.2

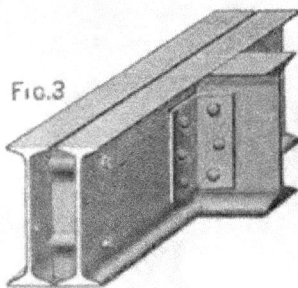

Fig.3

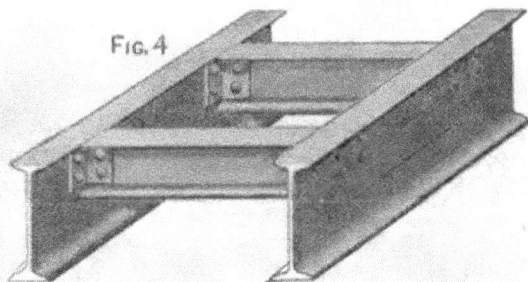

Fig.4

Fig.5

Fig.6

Fig.7

Cases frequently occur in which a column cannot be introduced into the building, and the girder must then be deepened and made strong enough to bear its load without such assistance. For this purpose girders are built of plate and angle irons combined in suitable form to resist the strains induced by the load in the several members, and of depths that vary to suit the special conditions of each case.

Fig. 8 shows the usual form adopted for plate girders. The ends should be further stiffened by vertical members, to resist the shearing strain on the web at the points of support, as shown on opposite page.

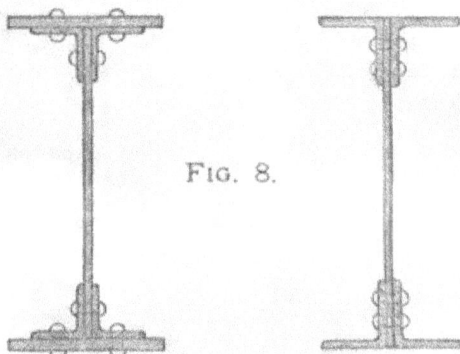

FIG. 8.

Box girders (as below) composed of a combination of plates with angle irons, are also frequently used, and may be built up in sections, varying according to architects' designs.

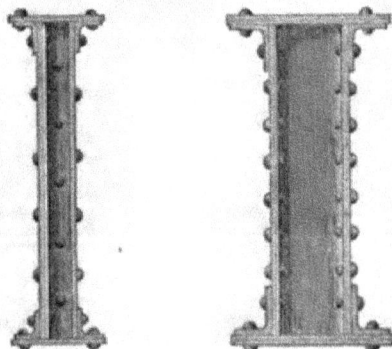

PLATE GIRDER.

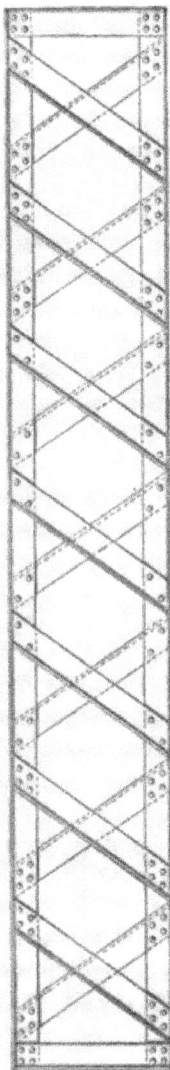

LATTICE GIRDER.

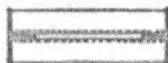

Between the joists the spaces are filled up with brick arches, resting on the lower flanges against cast-iron or brick skew-backs.

The bricks should be moulded with a slight taper to suit the arch, and be laid in place with as little mortar as possible. Above the arch the space is filled with grouting, in which wooden strips $2'' \times 1''$ are bedded for nailing the flooring to. The thrust of the arches is taken up by a series of tie-rods, placed in lines from 6 to 8 feet apart, and usually from $\frac{3}{4}$ to 1 inch in diameter, as shown in plan (Fig. 9), that run from beam to beam from one end of the building to the other, being anchored into each end wall with stout washers, an angle bar or channel serving as a wall-plate for distributing the strain produced by the thrust of the first arch.

Instead of the brick arches corrugated iron is sometimes used to fill in the spaces. It is placed on the lower flanges of the beams and filled in above with cement in place of brickwork.

The centres for turning the arches can be suspended by iron straps hooked on the lower flange, and detachable on one side so that the frames can be shifted from point to point as the work progresses. If a flush surface is preferred for the ceiling, it may be obtained by wedging strips of pine between the beams, and tacking the laths diagonally to the under side of these, finishing with a smooth and fair surface of plastering, and thus entirely concealing the iron-work above. Hollow brick, moulded especially for this class of work, have been used to some extent in the place of solid arching, with the object of diminishing the dead weight. The cost, however, is somewhat greater than solid bricks. Latterly, also, what are called flat arches, made of hollow bricks, have been introduced, the object being to secure a flat ceiling.

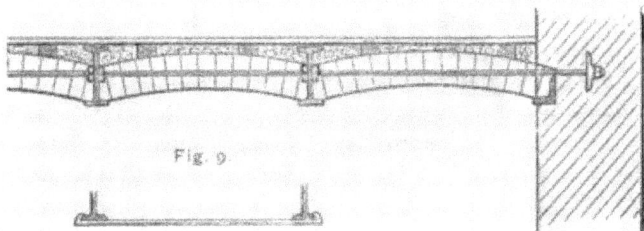

Fig. 9.

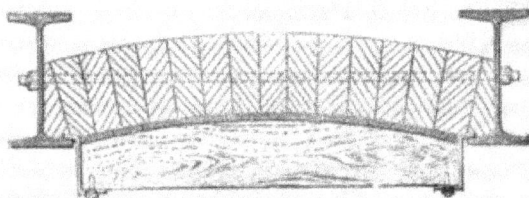

Fig. 10.

The use of hollow bricks and hollow composition blocks of a variety of shapes as a substitute for solid brick arches has become quite general, and illustrations of their useful application in the construction of fire-proof work are shown on the opposite page.

It is evident that the diminution of the dead load to be borne by the iron framing affords quite an advantage and permits of a more economical use of material.

The most effective method of accomplishing this result is to substitute hollow burnt-clay brick, or hollow concrete blocks, for the solid common bricks generally employed, thus reducing the dead weight of the arch by 40 to 50 per cent. The hollow brick and blocks may be used either in segmental or flat arches, according to whether a curved or flat ceiling is preferred.

Hollow blocks of burnt fire-clay, purposely made for use in flat arches, are manufactured in quantity in a number of places, and concrete blocks or artificial stone has also been employed with very satisfactory results. The voussoir blocks are cemented together with joints inclined to a common centre as in a segmental arch. The skew-backs of the flat arches take the form of the iron beams against which they rest, and each block keys with the adjacent one, no two joints being allowed to be parallel, as this would endanger the safety of a flat arch. The lower surfaces of the blocks descend about an inch below the flanges of the iron beams, and a thin tile is slipped into place to cover the iron for protection from fire. A coat of cement is then applied to the surface of the entire ceiling, and it is ready to receive any finishing decorative treatment that may be preferred. The upper level of the blocks may be carried up to the top of the iron beams, taking the place of the concrete filling sometimes employed. The iron beams will thus be entirely surrounded by the best known non-conductors of heat, brick or concrete, and will be fully protected from the action of flame, should the combustible contents of a room be accidentally burned.

For large spans a rib is formed in the hollow blocks following the curve of pressure, and this adds very materially

FIRE PROOF CONSTRUCTION
WITH IRON AND HOLLOW BRICK.

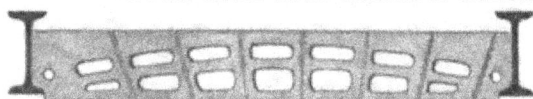

FLAT ARCH OF TEIL HOLLOW BLOCKS.

FLAT ARCH OF HOLLOW BRICK.

FLAT ARCH OF TEIL HOLLOW BLOCKS.

FLAT ARCH OF HOLLOW BRICK, ARCHED RIB.

HOLLOW BRICKS.

SKEW-BACKS.

Has sustained, without breaking, a load of 1291 lbs. per sq. ft.

FLAT ROOF BETWEEN IRON BEAMS.

POROUS LIGHT BRICK ARCHES AND BEAM PROTECTION.

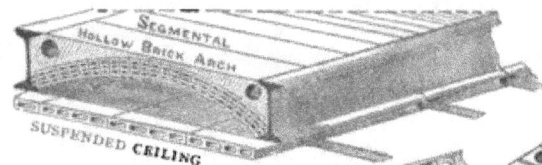

SEGMENTAL HOLLOW BRICK ARCH

SUSPENDED CEILING

ROOF OF HOLLOW BLOCKS.

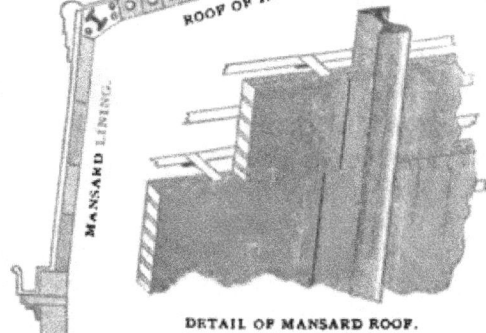

MANSARD LINING.

BLOCK FOR PARTITIONS OR MANSARD ROOFS.

DETAIL OF MANSARD ROOF.

PROTECTION FOR BOX GIRDERS.

to the strength of a flat arch formed of them. Such arches have frequently been tested with loads of one ton per square foot without failure, and their great strength, in combination with lightness, is of value and importance. But the blocks must be of first-class quality and skilfully placed by competent workmen to obtain the best results from them.

When segmental arches are preferred, hollow brick may, with advantage, be substituted for the ordinary solid bricks, diminishing the dead load to some extent. Suspended ceilings of hollow blocks 1½ to 2 inches thick are sometimes employed. The blocks are supported on bars of ⊥ and ∟ iron placed about 16 inches apart and hung from the floor beams by suitable hooks and clamps. The suspended ceiling is fire-proof in itself when coated with a covering of cement, and by means of the air space above it very thoroughly protects the floor beams from the effects of heat in the room below. Similar hollow blocks, well cemented together and bound with hoop iron about the flanges, are also used to protect box-girders from the effects of heat.

For making a finish inside the slating, and for lining Mansard roofs between the iron beams, hollow blocks 2 to 4 inches thick have been employed with excellent results. The blocks are usually cemented together and fastened to the purlins by small flat iron hooks, leaving a hollow space between the slating and the fire-proof hollow wall, the inner surface being smoothly plastered and finished.

Similar construction would be well adapted to vaults, domes, and the lining of refrigerator walls, where the non-conduction of heat is of importance. Rooms thus protected are dry and comfortable under any circumstances, being cool in summer and warm in winter. Hollow blocks are in very general use also for partitions in buildings, and when used in connection with floors of iron beams, protected by arches of the construction just described, they divide a building into a number of fire-proof compartments. If a fire originates in any one of these it is prevented from extending to the contents of the entire structure, and time is afforded for its easy extinction without risk of extensive damage by water or of injury to any part of the building itself.

PHŒNIX PATENT WROUGHT IRON COLUMNS,

AND

Method of Fire-Proofing and Preparing for Smooth Finish by Wight's Patent Process.

By the use of a non-conducting and incombustible casing Phœnix columns can be made thoroughly secure from the effects of expansion caused by fire in the combustible contents of rooms. They may, by the same means, be given any desired form and prepared for an exterior surface finish of cement. This cement finish may be in any desired color or may be highly polished to resemble marble. The process of protecting the columns consists in the use of terra-cotta blocks moulded to fit between the flanges of the segments, bedded in place with cement mortar, and secured by countersunk iron plates hooked over the rivet-heads of the columns. Fig. 2 is a perspective view of such a column, showing the various stages of completion.

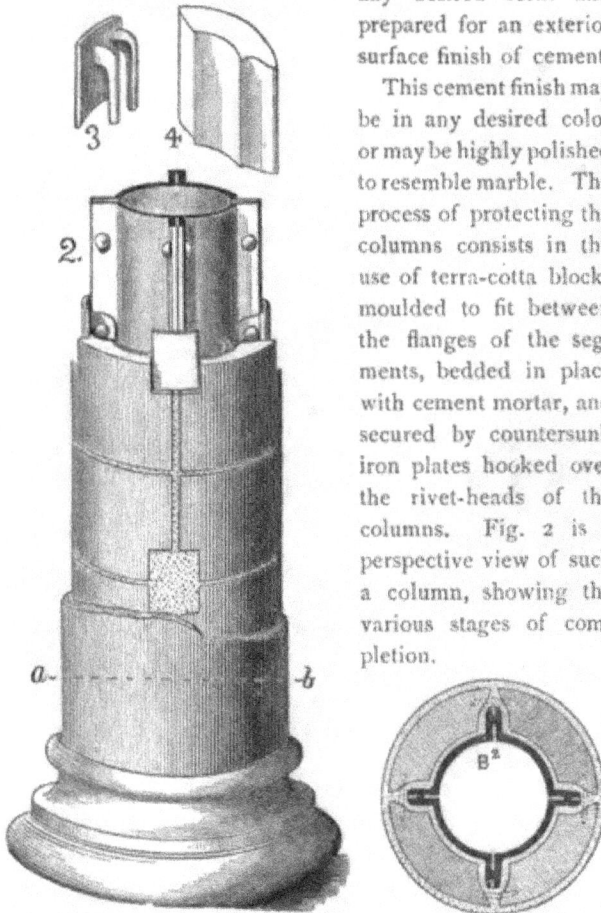

COLUMNS.

Wrought-iron columns are coming into more general use in the construction of buildings, both on account of the saving of space that they afford when compared with heavy walls of masonry, and because of the great loads that are now to be provided for in large fire-proof buildings. In the latter case cast-iron columns are generally more costly, and neither so safe nor so durable in the event of fire. The Phœnix column of wrought-iron segments, circular in section, provides the maximum of strength with the minimum of weight in the column itself.

To carry a given load, it requires the employment of the least amount of metal, and, on account of the simplicity of its construction, it is the cheapest as well as the best column in the market.

Whenever Phœnix columns are employed, the interior surfaces are thoroughly painted before the segments are riveted together. Such columns have been inspected after twenty years of service, and, although they had occupied the most exposed situations, they have been found uninjured by rust and with the paint still performing its duty as a protector. To determine the value of Phœnix columns under loads, a series of tests have been made at various times, the most noteworthy, probably, being those made on the Government machine at Watertown Arsenal, Massachusetts, in 1879, upon a set of full-sized Phœnix columns, of lengths ranging from 6 diameters to 42 diameters. Twenty C columns, each of about 12 square inches sectional area, were thus tested, and from these experiments the following formulæ have been deduced, which closely correspond with

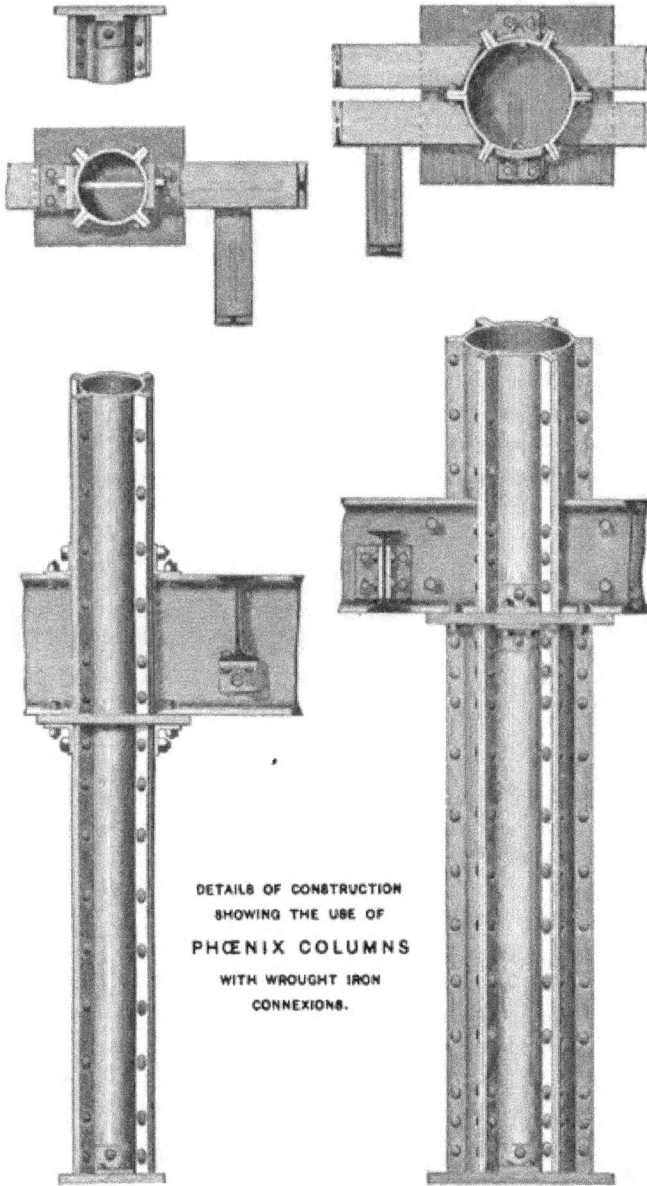

DETAILS OF CONSTRUCTION
SHOWING THE USE OF

PHŒNIX COLUMNS

WITH WROUGHT IRON
CONNEXIONS.

the actual results obtained, and show correctly the value of the form of the Phœnix column:

Formula for
Square-End Bearings.

$$\frac{P}{S} = \frac{42,000}{1 + \left(\frac{1}{50,000} \times \frac{l^2}{r^2}\right)}$$

Formula for
Pin-End Bearings.

$$\frac{P}{S} = \frac{42,000}{1 + \left(\frac{1}{30,000} \times \frac{l^2}{r^2}\right)}$$

In these formulæ the expression $\frac{P}{S}$ represents the $\frac{\text{total load in pounds}}{\text{sectional area in square inches}}$; or, in other words, the crushing strain per square inch of section. l is the length in feet between bearings, and r is the least radius of gyration. Applying these formulæ to the several patterns of segmental columns, the table of allowable working strains per square inch of section, shown below, has been prepared; the allowable working strains being, in each case, about one-fourth of the ultimate strength of the column.

ALLOWABLE
WORKING LOADS FOR PHŒNIX COLUMNS.
In Pounds per Square Inch of Sectional Area.
Square-End Bearings.

Length in Feet.	Col. A.	Col. B¹.	Col. B².	Col. C.	Col. E.	Col. G.
10	9323	9833	10,024	10,195	10,351	10,411
12	8885	9564	9,830	10,067	10,288	10,371
14	8420	9267	9,607	9,924	10,215	10,326
16	7943	8944	9,364	9,783	10,131	10,275
18	7463	8610	9,105	9,575	10,037	10,216
20	6997	8260	8,830	9,386	9,935	10,152
22	6526	7906	8,541	9,185	9,824	10,082
24	6090	7550	8,250	8,973	9,705	10,005
26	......	7201	7,955	8,755	9,580	9,926
28	......	6860	7,660	8,527	9,450	9,841
30	......	6527	7,366	8,297	9,314	9,750
32	......	......	7,075	8,070	9,170	9,654
34	......	......	......	7,837	9,021	9,555
36	......	......	......	7,604	8,870	9,441
38	......	......	......	7,375	8,717	9,341
40	......	......	......	7,147	8,561	9,235

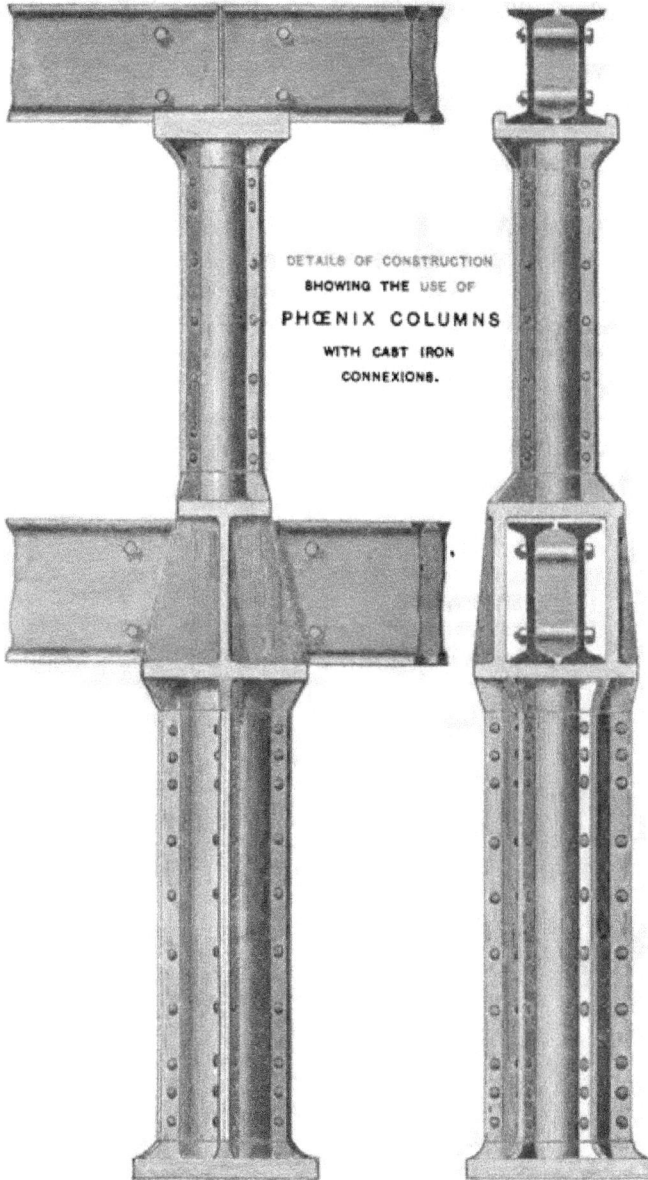

DETAILS OF CONSTRUCTION
SHOWING THE USE OF
PHŒNIX COLUMNS
WITH CAST IRON
CONNEXIONS.

TABLE OF DIMENSIONS OF PHŒNIX COLUMNS.

The dimensions given in the following table are subject to slight variations, which are unavoidable in rolling iron shapes.

The weights of columns given are those of the 4, 6, or 8 segments, of which they are composed. The *shanks* of the rivets used in joining the segments together only make up the quantity of metal removed in making the holes, but the *rivet-heads* add from 2 to 5 per cent. to the weights given. The rivets are spaced 3, 4, or 6 inches apart from centre to centre, and somewhat more closely at the ends than towards the centre of the column.

Any desired thickness between the minimum and maximum for any given size can be furnished. G columns have 8 segments, E columns 6 segments, C, B², B¹, and A have 4 segments.

Least Radius of Gyration equals D × .3636.

MARK	THICKNESS	DIAMETERS IN INS.			ONE COLUMN.			SIZE OF RIVETS.
		d Inside.	D Outside.	D¹ Over Flanges	Area of Cross Section Sq. Inches.	Weight per Foot in Pounds.	Least Radius of Gyration. Inches.	
A	$\frac{3}{16}$	$3\frac{5}{8}$	4	$6\frac{1}{16}$	3.8	12.6	1.45	$\frac{5}{8} \times 1\frac{1}{4}$
	$\frac{1}{4}$	"	$4\frac{1}{8}$	$6\frac{3}{16}$	4.8	16.0	1.50	$1\frac{3}{8}$
	$\frac{5}{16}$	"	$4\frac{1}{4}$	$6\frac{5}{16}$	5.8	19.3	1.55	$1\frac{1}{2}$
	"	"	$4\frac{3}{8}$	$6\frac{7}{16}$	6.8	22.6	1.59	$1\frac{5}{8}$
B¹	$\frac{1}{4}$	$4\frac{13}{16}$	$5\frac{1}{16}$	$8\frac{1}{16}$	6.4	21.3	1.92	$\frac{1}{2} \times 1\frac{5}{8}$
	$\frac{5}{16}$	"	$5\frac{3}{16}$	$8\frac{1}{8}$	7.8	26.0	1.96	$1\frac{3}{4}$
	$\frac{3}{8}$	"	$5\frac{5}{16}$	$8\frac{1}{4}$	9.2	30.6	2.02	$1\frac{7}{8}$
	"	"	$5\frac{7}{16}$	$8\frac{3}{8}$	10.6	35.3	2.07	$1\frac{1}{2}$
	"	"	$5\frac{9}{16}$	$8\frac{7}{16}$	12.0	40.0	2.11	$1\frac{7}{8}$
	"	"	$5\frac{11}{16}$	$8\frac{1}{2}$	13.4	44.6	2.16	2
	$\frac{9}{16}$	"	$6\frac{1}{16}$	$8\frac{5}{8}$	14.8	49.3	2.20	$2\frac{1}{8}$
B²	$\frac{1}{4}$	$5\frac{13}{16}$	$6\frac{7}{16}$	$9\frac{1}{4}$	7.4	24.6	2.34	$\frac{1}{2} \times 1\frac{5}{8}$
	$\frac{5}{16}$	"	$6\frac{9}{16}$	$9\frac{1}{4}$	9.0	30.0	2.39	$1\frac{3}{4}$
	"	"	$6\frac{11}{16}$	$9\frac{5}{8}$	10.6	35.3	2.43	$1\frac{7}{8}$
	$\frac{7}{16}$	"	$6\frac{13}{16}$	$9\frac{1}{2}$	12.2	40.6	2.48	$1\frac{7}{8}$
	"	"	$6\frac{15}{16}$	$9\frac{1}{2}$	13.8	46.0	2.52	$1\frac{7}{8}$
	$\frac{9}{16}$	"	$7\frac{1}{16}$	$9\frac{5}{8}$	15.4	51.3	2.57	2
	"	"	$7\frac{3}{16}$	$9\frac{11}{16}$	17.0	56.6	2.61	$2\frac{1}{8}$

MARK	THICKNESS	DIAMETERS IN INS. d Inside.	D Outside.	D1 Over Flanges.	ONE COLUMN. Area of Cross Section. Sq. Inches.	Weight per Foot in Pounds.	Least Radius of Gyration. Inches.	SIZE OF RIVETS.
C	1/4	7 7/16	7 11/16	11 9/16	10.0	33.3	2.80	5/8 × 1 1/4
	5/16	"	7 3/4	11 1/2	12.0	40.0	2.85	1 1/2
	3/8	"	7 13/16	11 1/2	14.0	46.6	2.90	2
	7/16	"	8 1/16	11 3/8	16.0	53.3	2.94	2 1/4
	1/2	"	8 3/16	11 3/8	18.0	60.0	2.98	2 1/4
	9/16	"	8 5/16	11 1/4	19.2	64.0	3.03	2 2/8
		"	8 7/16	12	21.2	70.6	3.08	3/4 × 2 2/8
	11/16	"	8 9/16	12 1/8	23.2	77.3	3.12	2 2/8
	3/4	"	8 11/16	12 1/8	25.2	84.0	3.16	2 2/8
	13/16	"	8 13/16	12 5/8	27.2	90.6	3.21	2 2/8
	7/8	"	8 15/16	12 7/8	29.2	97.3	3.26	3
	1	"	9 3/16	12 5/8	33.2	110.6	3.34	3 1/8
	1 1/8	"	9 7/16	12 7/8	37.2	124.0	3.43	3 1/4
	1 1/4	"	9 11/16	12 13/16	41.2	137.3	3.52	
E	1/4	11	11 1/8	15 7/16	16.8	56.	4.18	5/8 × 2
	5/16	"	11 1/4	15 9/16	19.2	64.	4.23	2 1/4
	3/8	"	11 1/2	15 1/2	21.6	72.	4.28	2 1/4
	7/16	"	11 7/8	15 3/8	24.0	80.	4.32	2 1/4
	1/2	"	12	15 1/2	26.4	88.	4.36	2 1/4
	9/16	"	12 1/4	16	28.8	96.	4.40	2 1/4
	5/8	"	12 3/8	16 1/4	31.8	106.	4.45	2 1/2
	11/16	"	12 3/8	16 3/8	34.8	116.	4.50	3/4 × 2 1/2
	3/4	"	12 1/2	16 5/16	37.8	126.	4.55	2 1/2
	13/16	"	12 5/8	16 7/8	40.8	136.	4.60	2 1/2
	7/8	"	12 5/8	16 5/8	43.8	146.	4.64	3
	1	"	13	16 3/4	49.8	166.	4.73	3
	1 1/8	"	13 1/4	17	55.8	186.	4.82	3 1/8
	1 1/4	"	13 1/2	17 3/16	61.8	206.	4.91	3 1/4
G	5/16	14 3/8	15	19 1/2	24.	80.0	5.45	5/8 × 2
	3/8	"	15 1/8	19 1/4	28.	93.3	5.50	2
	7/16	"	15 1/4	19 5/8	32.	106.6	5.55	2 1/4
	1/2	"	15 1/4	19 7/16	36.	120.0	5.59	2 1/4
	9/16	"	15 1/2	19 1/2	40.	133.3	5.63	2 1/4
	5/8	"	15 1/2	19 5/8	44.	146.6	5.68	2 1/4
	11/16	"	15 5/8	19 5/8	48.	160.0	5.72	3/4 × 2 1/2
	3/4	"	15 5/8	19 1/2	52.	173.3	5.77	2 1/2
	13/16	"	16	20	56.	186.6	5.82	2 1/2
	7/8	"	16 1/4	20 1/4	60.	200.0	5.87	2 1/2
	1	"	16 3/8	20 3/8	68.	226.6	5.95	3
	1 1/8	"	16 5/8	20 5/8	76.	253.3	6.04	3 1/8
	1 1/8	"	16 7/8	20 3/8	84.	280.0	6.14	3 1/4
	1	"	17 1/8	21	92.	306.6	6.23	3 3/8

ROOFS.

Iron trusses for rafters have been rapidly growing into favor with architects of late, owing in large measure to the combined lightness, strength, durability, and consequent economy of such structures. Various forms have been proposed for the trusses, some of the best known of which are here shown.

Figs. 11 and 15 are familiar illustrations. Fig. 12 shows the modification of the ordinary King and Queen truss as adapted to wrought iron, and Figs. 13 and 14 give examples of arched trusses that have been employed to cover depots and market-houses when a pleasing shape has been sought for the general outline of the building. For simplicity and economic arrangement of material, the design exhibited in Figs. 11 and 15 offers advantages over either of the other forms, and is most generally adopted in practice.

For the principals, T or I beams make very good rafters, and in light trusses T bars, or two channel bars $][$ either with or without a plate riveted to the upper flanges, answer every purpose. Struts may be made of light columns $\square$ A or B, of T bars, or of angle iron T, any of these forms affording great facility for attachment to the rafters.

For arched-roof trusses the details of construction are very similar to those described for peaked roofs; but as they are capable of great variety of treatment, the best illustrations that can be given of their forms will be by referring to Figs. 13 and 14—the highly ornamental and substantial roofs constructed by the Phœnix Iron Company for the market-house corner of Twelfth and Market Streets, Philadelphia, and for the station-shed at Altoona, on the Pennsylvania Railroad. These instances show the wide range of which the subject is susceptible.

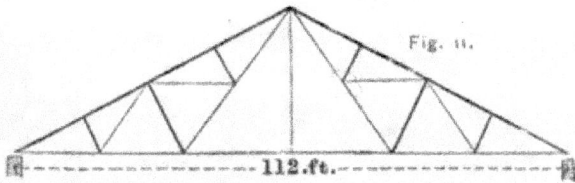

Fig. 11.

112.ft.

NEW MILL
PHŒNIX IRON WORKS.   ROCK ISLAND ARSENAL.

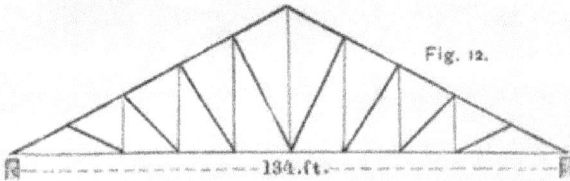

Fig. 12.

134.ft.

MASONIC TEMPLE, Philadelphia.

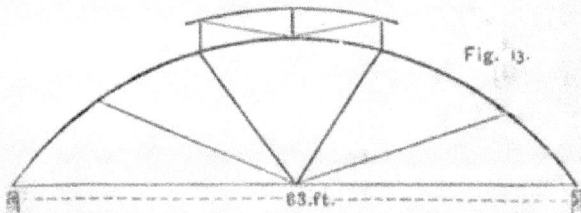

Fig. 13.

83.ft.

MARKET HOUSE, 12th and Market Sts., Philada.

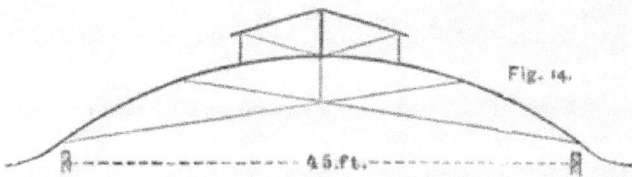

Fig. 14.

45.ft.

ALTOONA STATION, Penna. R.R.

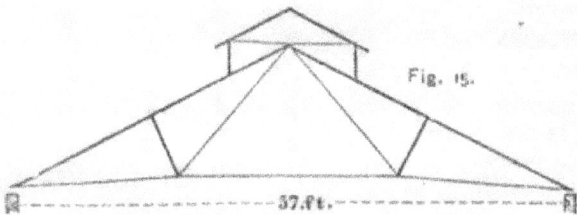

Fig. 15.

37.ft.

LEBANON FURNACE.

Ties may be of flat or round bars, attached by eyes and pins or screw ends. Care should be especially taken to properly proportion the dimensions of eyes and pins to the strains upon them. A very good and safe rule in practice is to make the diameter of the pin *from ¾ to ⅞ of the width of the bar in flats*, and 1¼ *times the diameter of the bar in rounds*, giving the eye a sectional area of 50 per cent. in excess of that of the bar. The *thickness* of *flat bars* should be at least *one-fourth of the width, in order* to secure good bearing surface on the pin, and the metal at the eyes should be as thick as the bars on which they are upset. Eyes are forged on the ends of flat or round bars by hydraulic pressure in suitably shaped dies, and, while the risk of a welded eye is thus avoided, a solid and well-formed eye is made from the iron of the bar itself. A similar process is adopted for enlarging the screw ends of long rods, so that when the screw is cut the diameter at the root of the thread is left a little larger than the body of the rod. Frequent trial with such rods has proven that they will pull apart in tension anywhere else but in the screw, the threads remaining perfect, and the nut turning freely after having been subjected to such a severe test. By this means the net section required in tension is made available with the least excess of material, and no more dead weight is put upon the structure than is actually required to carry the loads imposed.

The details of roof trusses vary to suit the character of the work and the sections of iron employed.

The heel of the rafter rests on the wall, either in a cast-iron skew-back fitted to the beam, and sloping to the angle required by the pitch of the roof, or between a couple of wrought angle-brackets riveted to the end of the rafter and resting on a wall-plate anchored to the wall. The struts are attached to the rafters by cast caps or by wrought strap-plates, and the joint at their feet is easily made either for pin or screw connexions. The peak is joined by wrought plates and bolts, the beams having been cut to the required angle.

Main rafters may be spaced from four to twenty feet apart, the spacing being regulated by the size of the purlin,

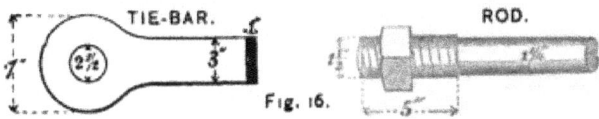

TIE-BAR.　　ROD.

Fig. 16.

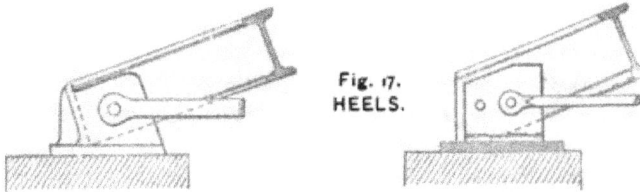

Fig. 17.
HEELS.

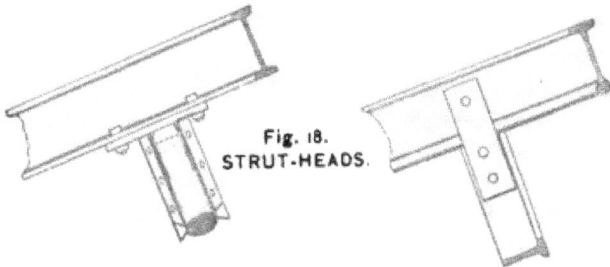

Fig. 18.
STRUT-HEADS.

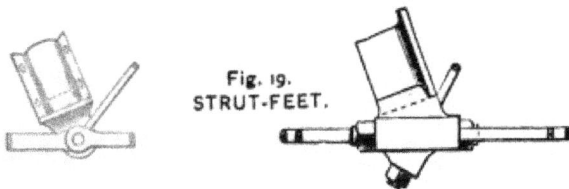

Fig. 19.
STRUT-FEET.

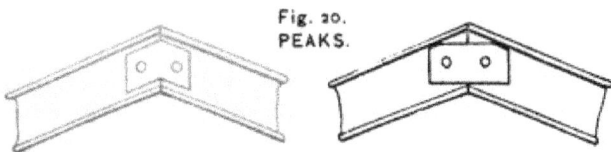

Fig. 20.
PEAKS.

and this again by the material used for covering. For slate on iron purlins a convenient spacing is about eight feet between centres of rafters, the angle-iron purlins being put at seven to fourteen inches apart, according to the size of the slate used, and notched at the ends into the flanges of the rafters. They are held in place by tie-rods that reach from rafter to rafter the entire length of the building, three or four rows of these rods being placed between peak and heel, at from six to eight feet intervals. On the iron purlins the slate may be laid directly and held down by copper or lead nails, clinched around the angle-bar, as shown in Fig. 21 ; or a netting of wire may be fastened to the purlins, and a layer of mortar spread on this, in which the slates are bedded. When greater intervals are used in spacing rafters, the purlins may be light beams fastened on top or against the sides of the principals with brackets, allowance always being made for longitudinal expansion of the iron by changes of temperature. On these purlins are fastened wooden jack-rafters carrying the sheathing-boards or laths, on which the metallic or slate covering is laid in the usual manner, or sheets of corrugated iron may be fastened from purlin to purlin, and the whole roof be entirely composed of iron.

When the rafters are spaced at such intervals as to cause too much deflexion in the purlins, they may be supported by a light beam, placed midway between the rafters and trussed transversely with posts and rods. These rods pass through the rafters, and have bevelled washers, screws, and nuts at each end for adjustment. By alternating thé trusses on either side of the rafter, and slightly increasing the length of the purlins above them, leaving all others with a little play in the notches, sufficient provision will be made for any alteration of length in the roof, due to changes of temperature.

When wooden purlins are employed they may be put between the rafters and held in place by tie-rods, or on top and fastened to the rafters by brackets; or hook-head spikes may be driven up into the purlin, the head of the spike hooking under the flange of the beam, spacing pieces of

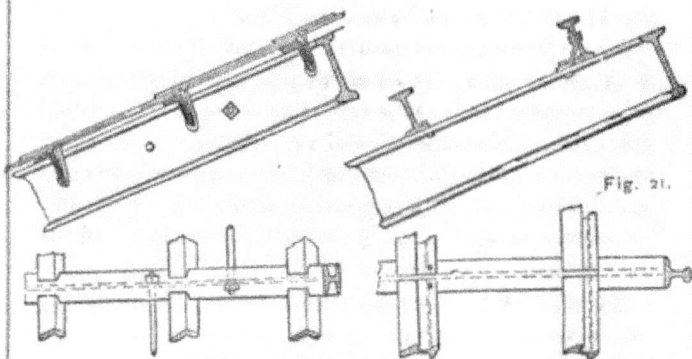

Fig. 21.

IRON PURLINS.

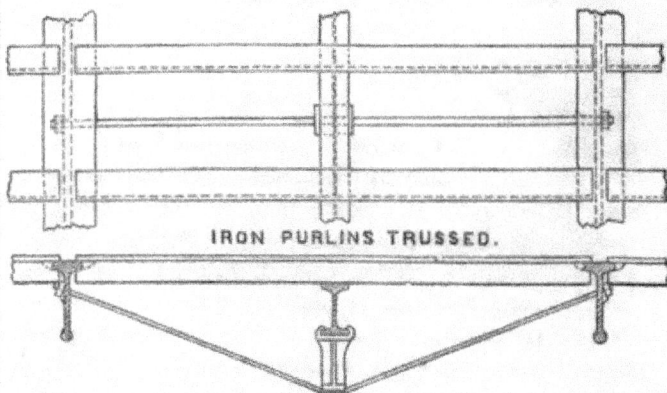

IRON PURLINS TRUSSED.

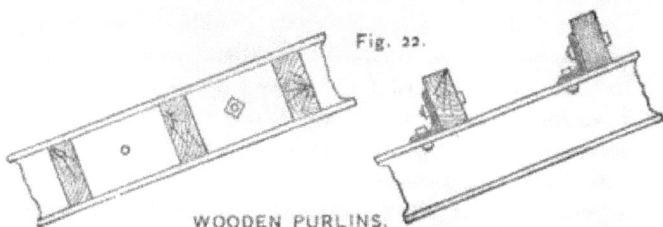

Fig. 22.

WOODEN PURLINS.

wood being laid on the top of the beam from purlin to purlin. The sheathing-boards and covering are then nailed
down on top of all in the usual manner.

When desired, ventilators or lanterns are added along
the ridge of the roof, as seen in Fig. 15, the attachments
being securely made to the rafters by wrought brackets and
bolts, and the bracing effected in a cheap and thorough
manner by two tie-rods that run from the peak of the rafter
to the angle between the post and rafter of the ventilator,
the covering material being attached as described for the
main rafters.

When it becomes desirable to suspend a ceiling from the
rafter, the tie-rods are replaced by a beam, and the ceiling
is attached to the lower flanges, curved T bars at the cornice serving to give any ornamental finish to the interior
that may suit the design of the architect.

For Mansard-roofs short additional beams are allowed to
project beyond the walls, and on these rest the feet of the
T bar or $\mathsf{[}$ bar framing, well fastened by wrought brackets and bolts. On the framing are secured the $1\frac{1}{2} \times \frac{3}{8}$
inch laths for attachment of the slate or metal covering,
and with a cornice of galvanized sheet iron perfect im-
munity from fire may be secured. This form of roof
work in wrought iron admits also of great scope for ornamental design, but from the amount of work required it
becomes rather more expensive than the less intricate com-
binations, and, as no two are alike in point of detail, it is
difficult to estimate the cost of construction. Curving,
shaping, and jointing the many pieces must be carefully
done to secure the close fitting that is requisite, and practical
experience in such work is of very great advantage to the
builder. (The roof of the new post-office in New York is
a very good illustration of the peculiarities of this class of
work.)

In Fig. 24 the purlins of angle-iron carry wooden strips,
to which are nailed the sheathing-boards and covering
material. A netting of wire may be used to attach the
plastering to the lower flanges of the tie-beams, or light

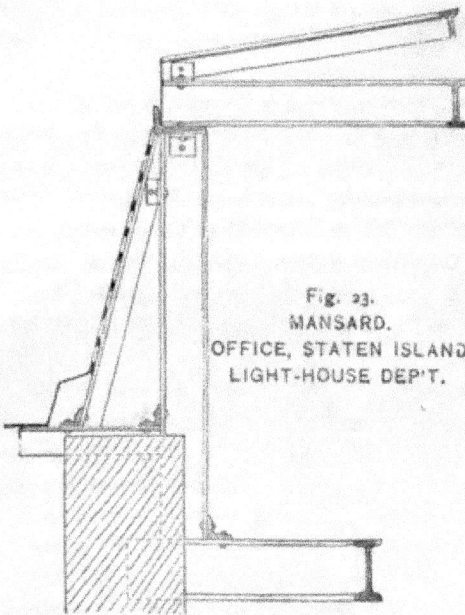

Fig. 23.
MANSARD.
OFFICE, STATEN ISLAND,
LIGHT-HOUSE DEP'T.

arches of tiles or hollow bricks may be turned on the lower flanges of smaller transverse beams as described for floors.

In roofs of wide span provision for expansion of the iron due to changes of *temperature* may be made by resting the skew-back of one end of the truss on a cast wall-plate, with rollers interposed to permit of the sliding of the heel without straining the wall, as in Fig. 25, but this precaution is not necessary in roofs of sixty feet span or less. Careful experiments have proved that an iron rod one hundred feet long will vary about $\frac{1}{15}$ of a foot for a change of temperature of 150 degrees Fahr., and as this is the greatest range to which iron beams and rods in a building would probably be subjected in this climate, compensation to that amount would be sufficient for all purposes. For sixty feet span the vibration of each wall would then be only $\frac{15}{1000}$ of a foot either way from the perpendicular, a variation so small and so gradually attained that there is no danger in imposing it upon the side walls by firmly fastening to them each heel of the rafter. Expansion is also provided against by fastening down one heel with wall-bolts and allowing the other to slide to and fro on the wall-plate without rollers, as shown in Fig. 17.

In estimating the strains on roofs the weight of the structure itself as well as the loads to be supported must be taken into account. Tredgold's assumption of the total maximum vertical load at forty pounds per square foot of horizontal surface is usually considered sufficiently high; but if a floor or ceiling is suspended to the tie-beam, or should the under side of the rafters be boarded and plastered, it is evident that these additional weights require more strength in the roof for their support.

For ordinary roofs of short span thirty pounds per square foot is quite enough, however, and for long spans, over sixty feet, thirty-five pounds will be sufficient to provide for, with the factors of safety in the material that are usually adopted. The stresses upon each member of the truss having been determined by any of the methods of calculation preferred, the sectional areas may be found by taking the safe tensile strength of good wrought iron at 10,000 pounds per square

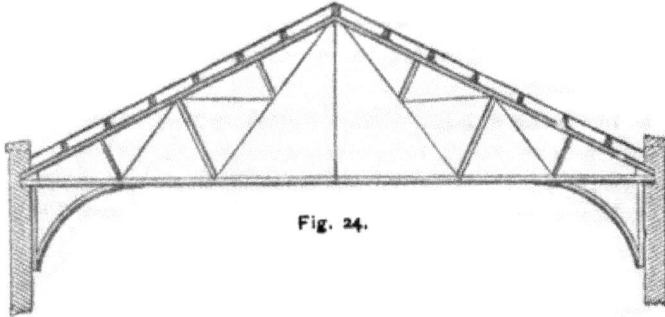

Fig. 24.

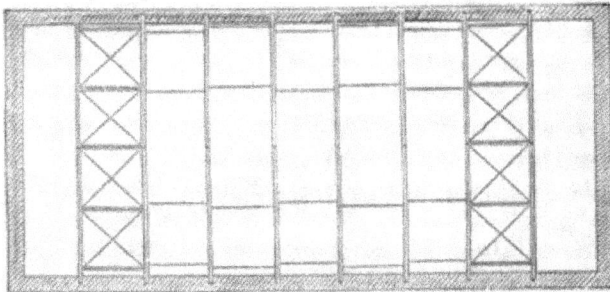

FRAMING and BRACING OF ROOF, Fig. 26.

Fig. 25.

inch, and the compressive resistance of beam or shape iron at from 6000 to 8000 pounds for the same unit of section.

It should be noted that the smaller or counterbrace rods ought to be made strong enough to resist strains induced by wind pressure on one side of the roof only,—the other half being unloaded.

Lateral braces, as in Fig. 26, should be provided in each end panel of straight roofs, as well to secure the roof during erection as to provide an abutment that will uphold the whole in case of fire or accident. From the panels so braced tie-rods run to each of the other rafters, and, with the purlins, unite the roof into a firm and compact whole. The gable walls are sometimes used to anchor the end rods into, but the method shown in the figure is that which is generally preferred.

A very economical combination of iron rafters with wrought-iron posts is shown in Fig. 27, this arrangement being well adapted for machine-shops, foundries, or other buildings in which it is desirable to cover a large area, and also to have an ample supply of light on the floor.

The posts on each side are placed from sixteen to twenty feet apart, and the heel of the intermediate rafter is supported by a trussed beam attached to the heads of the posts, the sheds on either side being covered by beams, trussed or untrussed, as the length of span may require. The skewback of the rafter and the cap of the post are cast in one piece, and all of the details of attachment between the parts are made in an equally simple and substantial manner. As a round-house for locomotives, or for many other purposes connected with railroad management, shops arranged on this plan commend themselves to the attention of engineers and master-mechanics, and for private establishments they have been found to answer their purpose admirably well, giving the maximum of surface covered at the minimum of first cost.

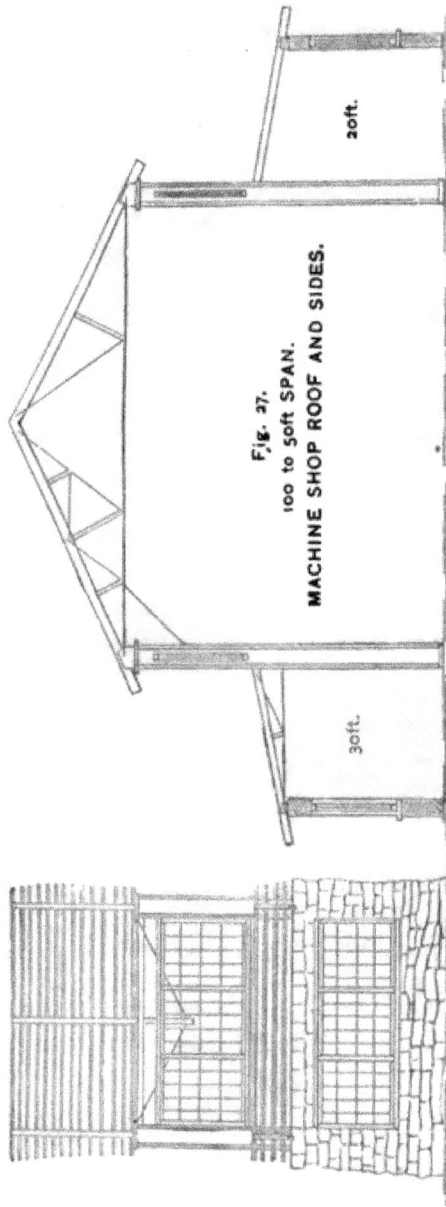

Fig. 27.
100 to 50ft SPAN.
MACHINE SHOP ROOF AND SIDES.

20ft.

30ft.

RECORD OF TESTS OF BEAMS.

TRANSVERSE STRENGTH.

As trustworthy data on which to base calculations for the efficiency of beams under transverse strain the tables given below are now published, having been the result of carefully conducted experiments on the part of the Phœnix Iron Company.

From these tables have been ascertained the coefficients for the safe load of each beam, so that it will be seen that dependence has not been placed merely on theoretical formulæ in assigning these values, but the truth of these formulæ has been demonstrated by the test of actual experiment.

7-inch Beam. 60 Lbs. per Yard. Area, 6 Sq. Inches. Clear Span, 21 Feet.				9-inch Beam. 87 Lbs. per Yard. Area, 8.7 Sq. Inches. Clear Span, 21 Feet.			
Centre Load, in Lbs.	Deflexion, Inches.	Increase, Inches.	Remarks.	Centre Load, in Lbs.	Deflexion, Inches.	Increase, Inches.	Remarks.
2,000	.468			2,000	.228		
3,000	.743	.275		4,000	.474	.246	
4,000	1.020	.277		6,000	.720	.246	
5,000	1.298	.278		8,000	.962	.242	
	.029 { Perm. set.	.280	Wt. rem'd.	10,000	1.201	.239	
6,000	1.578				.048 { Perm. set.	.231	Wt. rem'd.
	.030 { Perm. set.	.309	Wt. rem'd.	12,000	1.432		
7,000	1.887				.050 { Perm. set.	.148	Wt. rem'd.
	.060 { Perm. set.	.413	Wt. rem'd.	13,000	1.580		
8,000	2.300				.117 { Perm. set.	.283	Wt. rem'd.
	.183 { Perm. set.	1.240	Wt. rem'd.	14,000	1.863		
9,000	3.540				.269 { Perm. set.	1.393	Wt. rem'd.
9,500	5.298	1.758		16,000	3.256		
10,000 {			Beam sunk slowly, top flange yielding.	17,000	5.233	1.977 {	Side deflexion begins.
				17,500	5.602	.369 {	Beam yields slowly at this load.

9-inch Beam.
150 Lbs. per Yard. Area, 15 Sq. Inches. Clear Span, 14 Feet.

Centre Load, in Lbs.	Deflexion, Inches.	Increase, Inches.	Remarks.
5,608	.102		
6,720	.126	.024	
7,840	.148	.022	
8,960	.170	.022	
10,080	.192	.022	
11,200	.214	.022	
12,320	.239	.025	
13,440	.261	.022	
14,560	.287	.026	
15,680	.310	.023	
16,800	.336	.026	
17,920	.359	.023	
19,040	.382	.023	
20,160	.409	.027	
21,280	.435	.026	
22,400	.458	.023	
23,520	.487	.029	
24,640	.516	.029	
25,760	.543	.027	
26,880	.572	.029	
28,000	.600	.038	
29,120	.633	.033	load left stand 3/4 hour.
29,120	.682	.049	
	.082	Perm. set	Wt.rem.

15-inch Beam.
200 Lbs. per Yard. Area, 20 Sq. Inches. Clear Span, 14 Feet.

Centre Load, in Lbs.	Centre Load, Tons.	Deflexion, Inches.	Increase, Inches.
6,720	3	.048	
8,960	4	.060	.012
11,200	5	.073	.013
13,440	6	.090	.017
15,680	7	.105	.015
17,920	8	.120	.015
20,160	9	.134	.014
22,400	10	.148	.014
24,640	11	.161	.013
26,880	12	.178	.017
29,120	13	.191	.013
31,360	14	.206	.015
33,600	15	.222	.016
35,840	16	.234	.012
38,080	17	.246	.012
40,320	18	.258	.012
42,660	19	.271	.015
44,800	20	.287	.016
47,040	21	.305	.018

Weight removed. Permanent set, .016. After lapse of one hour the load of 15 tons was replaced, and caused a total deflexion of .222 inches as before.

12-inch Beam.
125 Lbs. per Yard. Area, 12½ Sq. Inches. Clear Span, 27 Feet.

Centre Load, in Lbs.	Deflexion, Inches.	Increase, Inches.
6,720	.691	
7,840	.821	.130
8,960	.948	.127
10,080	1.061	.113
11,200	1.186	.125
12,320	1.328	.142
13,340	1.466	.138
14,560	1.630	.164
15,680	1.800	.170
16,800	1.976	.176
17,920	2.228	.252
19,040	2.455	.227
20,160	2.742	.287
20,720	2.900	.158
20,720	2.965	.065

Last load left on 15 minutes. Deflexion increasing to 2.965.

15-inch Beam.
155 Lbs. per Yard. Area, 15½ Sq. Inches. Clear Span, 27 Feet.

Centre Load, in Lbs.	Deflexion, Inches.	Increase, Inches.
6,720	342	
7,840	.402	.060
8,960	.462	.060
10,080	.523	.061
11,200	.580	.057
12,320	.639	.059
13,440	.707	.068
14,560	.778	.071
15,680	.845	.067
16,800	.913	.068
17,920	.992	.079
19,040	1.063	.071
20,160	1.149	.086
22,400	1.309	.160
24,640	1.505	.196
25,760	1.603	.098

Load removed. Deflexion decreased to .261 permanent set after lapse of ½ hour.

RECORD OF TESTS OF PHŒNIX COLUMNS

Made with Hydraulic Press, 260 □" Piston Area.

SIZE.	Length.	Ratio of Length to Diameter.	Net Area, Square Inches.	Total Pressure on Piston, in Pounds.	Actual Ultimate Strength of Column per Square Inch.	Calculated Ultimate Strength by Gordon's Formula.	Shape of End Bearings.
May 3, 1873.							
B¹	8″	1.46	6.97	422 500	60 573	35 974	Flat.
B¹	8″	1.46	6.97	421 200	60 387	35 974	"
A	4″	0.92	5.62	370 500	65 867	35 990	"
A	4″	0.92	5.62	370 500	65 867	35 990	"
A	4″	1.01	2.92	166 400	56 889	36 000	"
A	4″	1.01	2.92	162 500	55 555	36 000	"
B¹	23.8′	53.5	5.84	176 800	30 274	18 430	"
B¹	24.′	53.6	5.95	97 590	16 387	7 457	Round.
C	23.3′	35.9	10.53	383 500	36 419	25 182	Flat.
C	22.8′	35.0	8.50	325 000	38 235	25 562	"
July 19, 1873.							
C	23.2′	34.5	13.31	436 800	32 742	25 774	"
C	23.2	34.5	12.85	455 000	35 408	25 774	"
June 2, 1875.							
C	27′	39.9	13.70	422 400	31 000	23 415	"
C	27	39.9	13.89	302 400	21 700	11 420	Round.
Aug. 5, 1875.							
C	28′	40.7	13.58	472 584	34 800	23 165	Flat.
C	28	40.7	13.58	497 028	36 600	23 165	"

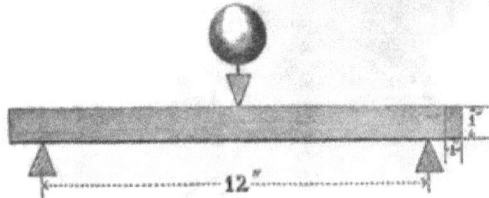

The breaking-load of a bar of wrought iron one inch square 12″ c. to c. of points of support is just 2240 pounds.

NOTES

CONCERNING SPECIFICATIONS OF QUALITY FOR IRON.

The tensile strength of iron is properly determined by ascertaining the load under which permanent set takes place, and the amount of stretch under the proof load, rather than from the ultimate load that causes the fracture of the bar. In other words, *the elastic limit* rather than the breaking strain should be regarded as the measure of quality in a bar, and working loads should be proportioned with reference to the elastic limit instead of to the so-called *ultimate strength.*

Tough, sinewy iron is what is required in a tension bar, and although a hard, unyielding iron may show greater ultimate strength under a gradually applied strain, yet it is not suitable for use under tension for the reason that a sudden shock may cause it to snap under a weight that it ought to carry with entire safety.

Good bar iron should be of uniform character and possess a limit of elasticity of not less than 25,000 pounds per square inch. The ultimate resistance of prepared test-bars having a sectional area of about one square inch for a length of 10 inches should be not less than 50,000 pounds per square inch when the test-bars have been prepared from full-sized bars having not more than 4 square inches of sectional area. For each additional square inch of full-sized bar area above 4 square inches a reduction of 500 pounds per square inch may be allowed down to a minimum ultimate resistance of 46,000 pounds. The amount of stretch under the breaking load should be not less than 15 per cent. in 10 inches of the test-bar.

Bars that are to be used in tension should stand, without cracking, a cold bending test to 90 degrees to a curvature the radius of which is about the thickness of the bar under test, and at least one third of the lot should stand bending to 180 degrees under the same conditions.

A round bar, one inch in diameter, should bend double, cold, without signs of fracture. A square bar of the same quality may show cracks on the edges under such a test.

Under a breaking pull the reduction of area should be not less than 25 per cent. of the original section.

The shape of a bar has much influence in determining the breaking-strain. The ultimate strength of round bars is, for this reason, considerably greater than that of flat bars, but in either case the elastic limit will be found to occur at about the same point for equally good qualities of iron.

Within the elastic limit the extension of iron may, for all practical purposes, be stated as follows :

Wrought iron, $\frac{1}{10000}$ of its length per ton per square inch.

Cast iron, $\frac{1}{30000}$ of its length per ton per square inch.

The compression of wrought iron within the limits of elasticity follows the same law, and the amount of shortening under pressure will be in direct proportion to the weight applied. But with cast iron the amount of compression does not follow a constant ratio, the compression per ton becoming greater with the increase of the weight. Thus, a cast iron bar, one square inch in section was compressed $\frac{1}{3000}$ of its length by a load of one ton; but under a load of 17 tons, instead of being compressed $\frac{17}{3000}$, it was compressed $\frac{28}{3000}$.

THE MODULUS OF ELASTICITY is a term used to designate such a *weight* as would extend a bar through a space equal to its original length, supposing the elasticity of the bar to be perfect. Or, the modulus of elasticity of any given material in *feet* is the height in feet of a column of this material, the weight of which would extend a bar of any determinate length through a space equal to this length. Thus, if one ton extends an inch bar of wrought iron one ten-thousandth of its length, it is evident that, upon the supposition that the bar is perfectly elastic, 10,000 tons would extend it to twice its original length. Hence, on this assumption, 10,000 tons, or 22,400,000 pounds, will be the modulus of elasticity of the wrought iron stated in *weight*. But an inch bar of wrought iron to weigh 22,400,000 pounds, at $3\frac{1}{3}$ pounds per foot, would be 6,720,000 feet long, and this would express the modulus of elasticity in *feet*.

The modulus of elasticity will, of course, vary according to the character of the material tested, being much higher in the better than it is in the lower grades of iron, but it forms a very useful and convenient standard of comparison in determining quality.

KIRKALDY'S CONCLUSIONS.

Mr. Kirkaldy sums up the results of his experimental inquiry in the following concluding observations, which the student should study carefully:

1. The breaking-strain does *not* indicate the quality, as hitherto assumed.

2. A *high* breaking-strain may be due to the iron being of superior quality, dense, fine, and moderately soft, or simply to its being very hard and unyielding.

3. A *low* breaking-strain may be due to looseness and coarseness in the texture, or to extreme softness, although very close and fine in quality.

4. The contraction of area at fracture, previously overlooked, forms an essential element in estimating the quality of specimens.

5. The respective merits of various specimens can be correctly ascertained by comparing the breaking-strain *jointly* with the contraction of area.

6. Inferior qualities show a much greater variation in the breaking-strain than superior.

7. Greater differences exist between small and large bars in coarse than in fine varieties.

8. The prevailing opinion of a rough bar being stronger than a turned one is erroneous.

9. Rolled bars are slightly hardened by being forged down.

10. The breaking-strain and contraction of area of iron plates are greater in the direction in which they are rolled than in a transverse direction.

22. Iron is less liable to snap the more it is worked and rolled.

33. The ratio of ultimate elongation may be greater in short than in long bars in some descriptions of iron, whilst in others the ratio is not affected by difference in the length.

44. Iron, like steel, is softened, and the breaking-strain reduced, by being heated and allowed to cool slowly.

54. A great variation exists in the strength of iron bars which have been cut and welded; whilst some bear almost as much as the uncut bar, the strength of others is reduced fully a third.

55. The welding of steel bars, owing to their being so easily burned by slightly overheating, is a difficult and uncertain operation.

56. Iron is injured by being brought to a white or welding heat, if not at the same time hammered or rolled.

57. The breaking-strain is considerably less when the strain is applied suddenly instead of gradually, though some have imagined that the reverse is the case.

61. The specific gravity is found generally to indicate pretty correctly the quality of specimens.

62. The density of iron is *decreased* by the process of wire-drawing, and by the similar process of cold rolling,* instead of *increased*, as previously imagined.

64. The density of iron is decreased by being drawn out under a tensile strain, instead of increased, as believed by some.

---

* NOTE.—The conclusion of Mr Kirkaldy in respect to cold rolling is undoubtedly true when the rolling amounts to wire-drawing; but when the compression of the surface by rolling diminishes the sectional area in greater proportion than it extends the bar, the result, according to the experience of the Pittsburgh manufacturers, is a slight increase in the density of the iron.

200. It must be abundantly evident from the facts which have been produced that the *breaking-strain* when taken alone gives a false impression of, instead of indicating, the real quality of the iron, as the experiments which have been instituted reveal the somewhat *startling* fact that frequently the inferior kinds of iron actually yield a higher result than the superior. The *reason* of this difference was shown to be due to the fact, that whilst the one quality retained its original area only very slightly decreased by the strain, the *other* was reduced to less than one-half. Now surely this variation, hitherto unaccountably *completely overlooked*, is of importance as indicating the relative hardness or softness of the material, and thus, it is submitted, forms an essential element in considering the *safe load* that can be practically applied in various structures. *It must be borne in mind* that although the softness of the material has the effect of lessening the amount of the breaking-strain, it has the very opposite effect as regards the *working-strain*. This holds good for two reasons: first, the softer the iron the less liable it is to snap; and second, fine or soft iron, being more uniform in quality, can be more *depended upon in practice*. Hence the load which this description of iron can suspend with safety may approach much more nearly the limit of its breaking-strain than can be attempted with the harder or coarser sorts, where a greater margin must necessarily be left.

202. As a necessary corollary to what we have just endeavored to establish, the writer now submits, in addition, that the *working-strain* should be in proportion to the breaking-strain per square inch of *fractured area*, and not to the breaking-strain per square inch of *original area* as heretofore. Some kinds of iron experimented on by the writer will sustain with safety more than double the load that others can suspend, especially in circumstances where the load is unsteady, and the structure exposed to concussions, as in a ship or railway bridge.

KIRKALDY'S RULE FOR COMPARING THE QUALITIES OF IRON:

The breaking-weight per square inch of the fractured area, instead of the breaking-weight or strain per square inch of the original area.

DIMINUTION OF TENACITY OF WROUGHT IRON

At High Temperatures.

EXPERIMENTS FRANKLIN INSTITUTE, 1839.

WALTER JOHNSON AND BENJAMIN REEVES, COM.

C.	Fahr.	Diminution per cent. of Max. Tenacity.	C.	Fahr.	Diminution per cent. of Max. Tenacity.
271°	520°	0.0738	500°	932°	0.3324
299		0.0869	508		0.3593
313		0.0899	554		0.4478
316		0.0964	599		0.5514
332	630	0.1047	624	1154	0.6000
350		0.1155	626		0.6011
378		0.1436	642		0.6352
389	732	0.1491	669		0.6622
390		0.1535	674	1245	0.6715
408		0.1589	708	1306	0.7001
410		0.1627			
440		0.2010			

The contraction of a wrought-iron rod in cooling is about equivalent to $\frac{1}{10000}$ of its length from a decrease of 15° Fahr., and the strain thus induced is about *one ton* for every square inch of sectional area in the bar.

For a rod of the lengths given below the contraction will be as follows:

Length of rod, in feet,	10	20	30	40	50	75	100	150
Contraction, in inches, for	15°	.012	.024	.036	.048	.060	.090	.120	.180
	100°	.080	.160	.240	.320	.400	.600	.800	1.200
	150°	.120	.240	.360	.480	.600	.900	1.200	1.800

Contraction and expansion being equal, the pressure per square inch induced by heating or cooling is as follows:

For temperatures varying by 15° Fahr.:

Variation,	15	30	45	60	75	105	120	150	degrees.
Pressure,	1	2	3	4	5	7	8	10	tons.

Stoney gives 8° C. = 14.4 Fahr. as equivalent to a pressure of one ton per square inch for wrought iron, and 15° C. = 27 Fahr. for cast iron.

LINEAR EXPANSION OF METALS.

	Between 0° and 100° C.	For 1° C.	For 1° Fahr.
Zinc	0.00294		
Lead	0.00284		
Tin	0.00222		
Copper, Yellow .	0.00188		
Copper, Red . .	0.00171		
Forged Iron* .	0.00122	.0000122	.00000677
Steel† . . .	0.00114	.0000114	.00000633
Cast Iron* . .	0.00111	.0000111	.00000616

For a change of 100° Fahr., a bar of iron 1475′ long will extend 1 foot. Similarly, a bar 100 feet long will extend .0678 foot, or .8136 inch.

According to the experiments of Du Long and Petit, we have the mean expansion of iron, copper, and platinum, between 0° and 100° C., and 0° and 300° C., as below :

	From 0° to 100° C.	0° to 300° C.
Iron	0.00180	0.00146
Copper	0.00171	0.00188
Platinum	0.00884	0.00918

The law for the expansion of iron, steel, and cast iron at very high temperatures, according to Rinman, is as follows :

	From 25° to 525° C. Red Heat=500° C.	For 1° C.	1° Fahr.
Iron	.00714	.0000143 =	.0000080
Steel	.01071	.0000214 =	.0000119
Cast Iron . .	.01250	.0000250 =	.0000139

	From 25° to 1300°. Nascent White=1275° C.		
Iron	.01250	.00000981 =	.00000545
Steel	.01787	.00001400 =	.00000777
Cast Iron . .	.02144	.00001680 =	.00000933

	From 500° to 1500°. Dull Red to White Heat=1000° C. Difference.		
Iron	.00535	.00000535 =	.0000030
Steel	.00714	.00000714 =	.0000040
Cast Iron . .	.00893	.00000893 =	.0000050

Ratio of Expansion in Hundred Parts, assuming Forge Iron to Expand between 0° and 100° C.=.00122.

	From 0° to 100°.	25° to 525°.	25° to 1300°.	500° to 1500°.
Iron . .	100 per ct.	117 per ct.	80 per ct.	44 per ct.
Steel .	93 "	175 "	114 "	58 "
Cast Iron	91 "	205 "	137 "	73 "

---

* Laplace and Lavoisier.   † Ramsden.

DIFFERENT COLORS OF IRON CAUSED BY HEAT.

POUILLET.

C.	FAHR.	COLOR.
210°	410°	Pale Yellow.
221	430	Dull Yellow.
256	493	Crimson.
261 370	502 680	Violet, Purple, and Dull Blue; between 261° C. to 370° C. it passes to Bright Blue, to Sea Green, and then disappears.
500	932	Commences to be covered with a light coating of oxide; loses a good deal of its hardness, becomes much more impressible to the hammer, and can be twisted with ease.
525	977	Becomes Nascent Red.
700	1292	Sombre Red.
800	1472	Nascent Cherry.
900	1657	Cherry.
1000	1832	Bright Cherry.
1100	2012	Dull Orange.
1200	2192	Bright Orange.
1300	2372	White.
1400	2552	Brilliant White—Welding Heat.
1500 1600	2732 2912	Dazzling White.

MELTING POINT OF METALS.

NAME.	FAHR.	FAHR.	AUTHORITY.
Platina	4593°		
Antimony	955	842	J. Lowthian Bell.
Bismuth	487	507	"
Tin (average)	475		
Lead "	622	620	"
Zinc	772	782	"
Cast Iron	2010	1922..2012 .. White. 2012..2192 .. Gray.	Pouillet.
Wrought Iron	2910	2733	Welding Heat. "
Steel	2370	2550	
Copper (average).	2174		

NOTES ON THE

WEIGHT AND COMPOSITION OF AIR.

1 cubic foot of air at $32°$ Fahr., under a pressure of 14.7 lbs. per square inch, weighs .080728 lb.
Therefore, 1000 cubic feet = 80.728 lbs.

1 cubic foot = 1.292 oz. . . . $\begin{cases} 23 \text{ per cent. Oxygen.} \\ 77 \text{ per cent. Nitrogen.} \end{cases}$

1 cubic foot of air contains . . . $\begin{cases} .29716 \text{ oz. Oxygen.} \\ .99484 \text{ oz. Nitrogen.} \end{cases}$

1.29200 total weight.

1 cubic foot of air contains . . . $\begin{cases} .0185725 \text{ lb. Oxygen.} \\ .0621555 \text{ lb. Nitrogen.} \end{cases}$

.080728 lb.

53.85 cubic feet of air contain . . $\begin{cases} 1.000 \text{ lbs. Oxygen.} \\ 3.347 \text{ lbs. Nitrogen.} \end{cases}$

4.347 lbs.

Carbonic acid = $C O_2$ = 22.
C = 6.   O = 8. O_2 = 16.   6 + 16 = 22.

For combustion to carbonic acid 1 lb. of coal requires $2\frac{3}{4}$ lbs. of oxygen, or 143.6 cubic feet of air, supposing all of the oxygen to combine with the coal. 280 to 300 cubic feet of air per pound of coal is the usual allowance for imperfect combustion.

11.59 lbs. of air for perfect combustion.
24 lbs. of air for imperfect combustion.

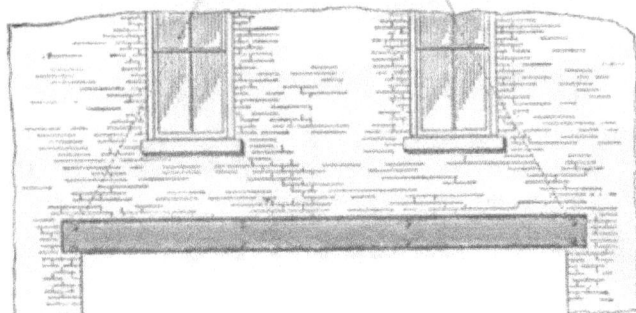

THE above cut illustrates a girder composed of two beams supporting a wall. During the construction a temporary prop should be placed beneath the girder after several courses of brick have been laid, and the prop should not be removed until the masonry is dry. This will prevent undue deflexion of the girder.

The girder should be of sufficient strength to sustain the entire weight of the wall between perpendicular lines above the span to a height corresponding to the apex of the dotted lines.

Assuming the weight of a cubic foot of brick wall to be 112 pounds, a superficial square foot of 9 inch wall will weigh 84 pounds, of 13 inch wall 121 pounds, and of 18 inch wall 168 pounds, and the following table specifies suitable beams for use as girders over the several spans named.

PROPER SIZES OF BEAMS TO USE AS GIRDERS FOR SUPPORTING WALLS.

SPAN.	13″ Wall.	SPAN.	13″ Wall.
Feet.		*Feet.*	
8 to 10	2—6″ 40 lbs.	18 to 20	2—10½″ 90 lbs.
10 to 12	2—7″ 55 lbs.	20 to 22	2—12″ 96 lbs.
12 to 14	2—8″ 65 lbs.	22 to 24	2—12″ 125 lbs.
14 to 16	2—9″ 70 lbs.	24 to 26	2—15″ 150 lbs.
16 to 18	2—9″ 84 lbs.	26 to 28	2—15″ 200 lbs.

TABLES

—OF—

Weights and Measures

WEIGHT OF FLAT BAR IRON.

PER FOOT.

Width in inches	THICKNESS, IN INCHES.											
	$\frac{1}{16}$	$\frac{1}{8}$	$\frac{3}{16}$	$\frac{1}{4}$	$\frac{5}{16}$	$\frac{3}{8}$	$\frac{7}{16}$	$\frac{1}{2}$	$\frac{5}{8}$	$\frac{3}{4}$	$\frac{7}{8}$	1
	lbs.	lbs.	lbs.	lbs.	lbs.	lbs.	lbs.	lbs.	lbs.	lbs.	lbs.	lbs.
1	.21	.42	.63	.84	1.05	1.26	1.47	1.68	2.11	2.53	2.95	3.37
1⅛	.24	.47	.71	.95	1.18	1.42	1.66	1.90	2.37	2.84	3.32	3.79
1¼	.26	.53	.79	1.05	1.32	1.58	1.84	2.11	2.63	3.16	3.68	4.21
1⅜	.29	.58	.87	1.16	1.45	1.74	2.03	2.32	2.89	3.47	4.05	4.63
1½	.32	.63	.95	1.26	1.58	1.90	2.21	2.53	3.16	3.79	4.42	5.05
1⅝	.34	.68	1.03	1.37	1.71	2.05	2.39	2.74	3.42	4.11	4.79	5.47
1¾	.37	.74	1.11	1.47	1.84	2.21	2.58	2.95	3.68	4.42	5.16	5.89
1⅞	.40	.79	1.18	1.58	1.97	2.37	2.76	3.16	3.95	4.74	5.53	6.32
2	.42	.84	1.26	1.68	2.11	2.53	2.95	3.37	4.21	5.05	5.89	6.74
2⅛	.45	.90	1.34	1.79	2.24	2.68	3.13	3.58	4.47	5.37	6.26	7.16
2¼	.47	.95	1.42	1.90	2.37	2.84	3.32	3.79	4.74	5.68	6.63	7.58
2⅜	.50	1.00	1.50	2.00	2.50	3.00	3.50	4.00	5.00	6.00	7.00	8.00
2½	.53	1.05	1.58	2.11	2.63	3.16	3.68	4.21	5.26	6.32	7.37	8.42
2⅝	.55	1.11	1.66	2.21	2.76	3.32	3.87	4.42	5.53	6.63	7.74	8.84
2¾	.58	1.16	1.74	2.32	2.89	3.47	4.05	4.63	5.79	6.95	8.10	9.26
2⅞	.61	1.21	1.82	2.42	3.03	3.63	4.24	4.84	6.05	7.26	8.47	9.63
3	.63	1.26	1.90	2.53	3.16	3.79	4.42	5.05	6.32	7.58	8.84	10.10
3¼	.68	1.37	2.05	2.74	3.42	4.11	4.79	5.47	6.84	8.21	9.58	10.95
3½	.74	1.47	2.21	2.95	3.68	4.42	5.16	5.89	7.37	8.84	10.32	11.79
3¾	.79	1.58	2.37	3.16	3.95	4.74	5.53	6.32	7.89	9.47	11.05	12.63
4	.84	1.68	2.53	3.37	4.21	5.05	5.89	6.74	8.42	10.10	11.79	13.47
4¼	.90	1.79	2.68	3.58	4.47	5.37	6.26	7.16	8.95	10.74	12.53	14.31
4½	.95	1.90	2.84	3.79	4.74	5.68	6.63	7.58	9.47	11.38	13.26	15.16
4¾	1.00	2.00	3.00	4.00	5.00	6.00	7.00	8.00	10.00	12.00	14.00	16.00
5	1.05	2.11	3.16	4.21	5.26	6.32	7.37	8.42	10.53	12.63	14.74	16.84
5¼	1.11	2.21	3.32	4.42	5.53	6.63	7.74	8.84	11.05	13.26	15.47	17.68
5½	1.16	2.32	3.47	4.63	5.79	6.95	8.10	9.26	11.58	13.89	16.21	18.52

WEIGHT OF FLAT BAR IRON.

PER FOOT.

Width in inches.	THICKNESS, IN INCHES.											
	1/16	1/8	3/16	1/4	5/16	3/8	7/16	1/2	5/8	3/4	7/8	1
	lbs.	lbs.	lbs.	lbs.	lbs.	lbs.	lbs.	lbs.	lbs.	lbs.	lbs.	lbs.
5¾	1.21	2.42	3.63	4.84	6.05	7.26	8.47	9.68	12.10	14.53	16.95	19.37
6	1.26	2.53	3.79	5.05	6.32	7.58	8.84	10.10	12.63	15.16	17.68	20.21
6¼	1.31	2.63	3.95	5.27	6.58	7.90	9.21	10.53	13.16	15.79	18.42	21.05
6½	1.36	2.73	4.10	5.47	6.84	8.21	9.58	10.94	13.68	16.42	19.16	21.88
6¾	1.42	2.84	4.26	5.69	7.10	8.53	9.95	11.36	14.21	17.05	19.90	22.73
7	1.47	2.94	4.42	5.90	7.36	8.84	10.32	11.79	14.74	17.68	20.64	23.58
7¼	1.53	3.05	4.58	6.11	7.63	9.16	10.68	12.21	15.26	18.32	21.37	24.42
7½	1.58	3.16	4.74	6.32	7.90	9.48	11.06	12.64	15.78	18.94	22.11	25.28
7¾	1.63	3.26	4.90	6.53	8.16	9.79	11.42	13.06	16.31	19.57	22.84	26.12
8	1.68	3.36	5.05	6.74	8.42	10.10	11.78	13.48	16.84	20.20	23.58	26.94
8¼	1.74	3.47	5.21	6.95	8.68	10.42	12.16	13.89	17.37	20.84	24.32	27.79
8½	1.79	3.58	5.36	7.16	8.94	10.74	12.52	14.32	17.90	21.48	25.06	28.63
8¾	1.84	3.68	5.53	7.37	9.21	11.05	12.89	14.74	18.42	22.10	25.79	29.47
9	1.90	3.79	5.68	7.58	9.48	11.36	13.26	15.16	18.95	22.75	26.52	30.32
9¼	1.95	3.90	5.84	7.79	9.74	11.68	13.63	15.58	19.47	23.38	27.26	31.16
9½	2.00	4.00	6.00	8.00	10.00	12.00	14.00	16.00	20.00	24.00	28.00	32.00
9¾	2.05	4.11	6.16	8.21	10.26	12.32	14.37	16.42	20.53	24.63	28.74	32.84
10	2.10	4.21	6.32	8.42	10.52	12.64	14.74	16.84	21.05	25.26	29.48	33.68
10¼	2.16	4.32	6.48	8.63	10.79	12.95	15.11	17.26	21.58	25.89	30.21	34.52
10½	2.21	4.41	6.64	8.84	11.05	13.26	15.48	17.68	22.10	26.52	30.95	35.36
10¾	2.26	4.53	6.79	9.05	11.32	13.58	15.84	18.10	22.63	27.16	31.68	36.21
11	2.32	4.64	6.95	9.26	11.58	13.90	16.21	18.52	23.16	27.78	32.42	37.04
11¼	2.37	4.74	7.11	9.47	11.85	14.21	16.58	18.94	23.68	28.42	33.15	37.89
11½	2.42	4.84	7.26	9.68	12.10	14.52	16.94	19.36	24.20	29.06	33.90	38.74
11¾	2.47	4.94	7.42	9.89	12.37	14.84	17.31	19.78	24.73	29.69	34.63	39.56
12	2.52	5.05	7.58	10.10	12.64	15.16	17.68	20.20	25.26	30.32	35.36	40.40

WEIGHT OF WROUGHT IRON.

Thickness or Diam. in Dec'ls, Inches.	Wt. of a Foot.	Wt. of a Sq. Foot, Lbs.	Wt. per Foot Sq. Bar, Lbs.	Wt. per Foot Round Bar, Lbs.
1/32	.0026	1.263	.0033	.0026
1/16	.0052	2.526	.0132	.0104
3/32	.0078	3.789	.0296	.0233
1/8	.0104	5.052	.0526	.0414
5/32	.0130	6.315	.0823	.0646
3/16	.0156	7.578	.1184	.0930
7/32	.0182	8.841	.1612	.1266
1/4	0208	10.10	.2105	.1653
9/32	.0234	11.37	.2665	.2093
5/16	.0260	12.63	.3290	.2583
11/32	.0287	13.89	.3980	.3126
3/8	.0313	15.16	.4736	.3720
13/32	.0339	16.42	.5558	.4365
7/16	.0365	17.68	.6446	.5063
15/32	.0391	18.95	.7400	.5813
1/2	.0417	20.21	.8420	.6613
9/16	.0469	22.73	1.066	.8370
5/8	.0521	25.26	1.316	1.033
11/16	.0573	27.79	1.592	1.250
3/4	.0625	30.31	1.895	1.488
13/16	.0677	32.84	2.223	1.746
7/8	.0729	35.37	2.579	2.025
15/16	.0781	37.89	2.960	2.325
1	.0833	40.42	3.368	2.645
1 1/16	.0885	42.94	3.803	2.986
1 1/8	.0938	45.47	4.263	3.348
1 3/16	.0990	48.00	4.750	3.730
1 1/4	.1042	50.52	5.263	4.133
1 5/16	.1094	53.05	5.802	4.557
1 3/8	.1146	55.57	6.368	5.001
1 7/16	.1198	58.10	6.960	5.466
1 1/2	.1250	60.63	7.578	5.952
1 5/8	.1354	65.68	8.893	6.985
1 3/4	.1458	70.73	10.31	8.101
1 7/8	.1563	75.78	11.84	9.300
2	.1667	80.83	13.47	10.58
2 1/8	.1771	85.89	15.21	11.95
2 1/4	.1875	90.94	17.05	13.39
2 3/8	.1979	95.99	19.00	14.92
2 1/2	.2083	101.0	21.05	16.53
2 5/8	.2188	106.1	23.21	18.23
2 3/4	.2292	111.2	25.47	20.01
2 7/8	.2396	116.2	27.84	21.87
3	.2500	121.3	30.31	23.81

WEIGHT OF WROUGHT IRON.

Thickness or Diam. in Dec'ls. Inches.	Wt. of a Sq. of a Foot.	Wt. of a Sq. Foot, Lbs.	Wt. per Foot Sq. Bar, Lbs.	Wt. per Foot Round Bar, Lbs.
3½	.2604	126.3	32.89	25.83
	.2708	131.4	35.57	27.94
	.2813	136.4	38.37	30.13
	.2917	141.5	41.26	32.41
	.3021	146.5	44.26	34.76
	.3125	151.6	47.37	37.20
	.3229	156.6	50.57	39.72
4	.3333	161.7	53.89	42.33
	.3438	166.7	57.31	45.01
	.3542	171.8	60.84	47.78
	.3646	176.8	64.47	50.63
	.3750	181.9	68.20	53.57
	.3854	186.9	72.05	56.59
	.3958	192.0	75.99	59.69
	.4063	197.0	80.05	62.87
5	.4167	202.1	84.20	66.13
	.4271	207.1	88.47	69.48
	.4375	212.2	92.83	72.91
	.4479	217.2	97.31	76.43
	.4583	222.3	101.9	80.02
	.4688	227.3	106.6	83.70
	.4792	232.4	111.4	87.46
	.4896	237.5	116.3	91.31
6	.5000	242.5	121.3	95.23
	.5208	252.6	131.6	103.3
	.5417	262.7	142.3	111.8
	.5625	272.8	153.5	120.5
7	.5833	282.9	165.0	129.6
	.6042	293.0	177.0	139.0
	.6250	303.1	189.5	148.8
	.6458	313.2	202.3	158.9
8	.6667	323.3	215.6	169.3
	.6875	333.4	229.3	180.1
	.7083	343.5	243.4	191.1
	.7292	353.6	247.9	202.5
9	.7500	363.8	272.8	214.3
	.7708	373.9	288.2	226.3
	.7917	384.0	304.0	238.7
	.8125	394.1	320.2	251.5
10	.8333	404.2	336.8	264.5
	.8750	424.4	371.3	291.6
11	.9167	444.8	407.5	320.1
	.9583	464.6	445.4	349.8
12	1 Foot.	485.	485.	380.9

GENERAL RULES

FOR DETERMINING

THE WEIGHT OF ANY PIECE OF WROUGHT IRON.

One cubic foot of wrought iron . . . . . $= 480$ lbs.

One square foot, one inch thick . . . $= \frac{480}{12} =$ 40 lbs.

One **square inch,** one foot long . . . $= \frac{40}{12} =$ $3\frac{1}{3}$ lbs.

One square inch, one yard long . . $= 3\frac{1}{3} \times 3 =$ 10 lbs.

Hence it **appears** that the weight of any piece of wrought **iron in pounds per yard** is equal to **10 times its area in square inches.**

Example.—The area of a bar $3'' \times 1'' = 3$ square inches, and its weight is 30 lbs. per yard.

For round iron the weight per **foot may be found** by taking the diameter in quarter inches, squaring it, and dividing by 6.

Example.—What is **the** weight of **2''** round iron?

$$2'' = 8 \text{ quarter inches.} \quad 8^2 = 64.$$

$$\tfrac{64}{6} = 10\tfrac{2}{3} \text{ lbs. per foot of } 2'' \text{ round.}$$

Example.—What is the weight of $\frac{3}{4}''$ round iron?

$$\tfrac{3}{4}'' = 3 \text{ quarter inches.} \quad 3^2 = 9.$$

$$\tfrac{9}{6} = 1\tfrac{1}{2} \text{ lbs. per foot of } \tfrac{3}{4}'' \text{ round.}$$

The above rules are highly convenient, and enable mental calculations of weight to be quickly obtained with accuracy.

CAST-IRON PIPE.

WEIGHT OF A LINEAL FOOT.

Bore, in Inches.	THICKNESS OF METAL, IN INCHES.								
	$\frac{1}{4}$	$\frac{3}{8}$	$\frac{1}{2}$	$\frac{5}{8}$	$\frac{3}{4}$	$\frac{7}{8}$	1	$1\frac{1}{8}$	$1\frac{1}{4}$
	lbs.	lbs.	lbs.	lbs.	lbs.	lbs.	lbs.	lbs.	lbs.
2	5.5	8.7	12.3	16.1	20.3	24.7	29.5	34.5	39.9
2½	6.8	10.6	14.7	19.2	24.0	29.0	34.4	40.0	46.0
3	7.9	12.4	17.2	22.2	27.6	32.3	39.3	45.6	52.2
3½	9.2	14.3	19.6	25.3	31.3	37.6	44.2	51.0	58.3
4	10.4	16.1	22.1	28.4	35.0	41.9	49.1	56.6	64.4
4½	11.7	18.0	24.5	31.5	38.7	46.2	54.0	62.1	70.6
5	12.9	19.8	27.0	34.5	42.3	50.5	59.9	67.7	76.7
5½	14.1	21.6	29.5	37.6	46.0	54.8	63.8	73.2	82.9
6	15.3	23.5	31.9	40.7	49.7	59.1	68.7	78.7	89.0
7	17.8	27.2	36.9	46.8	57.1	67.7	78.5	89.8	101.
8	20.3	30.8	41.7	52.9	64.4	76.2	88.4	101.	114.
9	22.7	34.5	46.6	59.1	71.8	84.8	98.2	112.	126.
10	25.2	38.2	51.5	65.2	79.2	93.4	108.	123.	138.
11	27.6	41.9	56.5	71.3	86.5	102.	118.	134.	150.
12	30.1	45.6	61.4	77.5	93.9	111.	128.	145.	163.
13	32.5	49.2	66.3	83.6	101.	119.	138.	156.	175.
14	35.0	52.9	71.2	89.7	109.	128.	147.	167.	187.
15	37.4	56.6	76.1	95.9	116.	136.	157.	178.	199.
16	39.1	60.3	81.0	102.	123.	145.	167.	189.	212.
18	44.8	67.7	90.9	114.	138.	162.	187.	211.	236.
20	49.7	73.2	101.	127.	153.	179.	206.	233.	261.
22	54.6	82.6	111.	139.	168.	197.	226.	255.	285.
24	59.6	89.9	120.	151.	182.	214.	245.	278.	310.
26	64.5	97.3	131.	164.	198.	231.	266.	300.	335.
28	69.4	105.	140.	176.	212.	249.	286.	323.	360.
30	74.2	112.	150.	188.	227.	266.	305.	345.	384.

NOTE.—For each joint, add a foot to length of pipe.

GALVANIZED AND BLACK IRON.

Weight in Pounds per Square Foot of Galvanized Sheet Iron, both Flat and Corrugated.

The numbers and thicknesses are those of the iron before it is galvanized. When a flat sheet (the ordinary size of which is from 2 to $2\frac{1}{2}$ feet in width, by 6 to 8 feet in length) is converted into a corrugated one, with corrugations 5 inches wide from centre to centre, and about an inch deep (the common sizes), its width is thereby reduced about $\frac{1}{10}$th part, or from 30 to 27 inches; and consequently the weight per square foot of area covered is increased about $\frac{1}{9}$th part. When the corrugated sheets are laid upon a roof, the overlapping of about $2\frac{1}{2}$ inches along their sides and of 4 inches along their ends diminishes the covered area about $\frac{1}{9}$th part more; making their weight per square foot of roof about $\frac{1}{9}$th part greater than before. Or the weight of corrugated iron per square foot in place on a roof is about $\frac{1}{5}$ greater than that of the flat sheets of above sizes of which it is made.

No. B. W. Gauge.	BLACK.				GALVANIZED.			
	Flat.		Corrugated.		Flat.		Corrugated.	
	Lbs.	On Roof.	Lbs.	On Roof.	Lbs.	On Roof.	Lbs.	On Roof.
30	.48	.56	.53	.62	.71	.83	.79	.91
29	.52	.61	.58	.68	.75	.87	.83	.97
28	.56	.67	.62	.73	.81	.94	.90	1.05
27	.64	.75	.71	.83	.87	1.01	97	1.13
26	.72	.84	.80	93	.94	1.09	1.04	1.21
25	.80	.93	.89	1.04	1.00	1.17	1.11	1.29
24	.88	1.03	98	1.14	1.06	1.24	1.18	1.37
23	1.00	1.17	1.11	1.29	1.19	1.39	1.32	1.54
22	1.12	1.31	1.24	1.45	1.31	1.53	1.47	1.71
21	1.28	1.49	1.43	1.67	1.50	1.75	1.67	1 95
20	1.40	1.63	1.56	1.82	1.75	2.03	1.94	2.26
19	1.69	1 97	1.87	2.18	1.94	2.26	2 15	2.51
18	1.96	2 29	2.18	2.54	2.37	2 76	2.63	3.07
17	2.33	2 72	2.59	3.02	2.69	3 13	2 99	3.49
16	2.60	3.03	2.89	3.37	3.00	3.50	3 33	3.88
15	2.89	3.37	3.21	3.74	3.30	3 85	3.67	4.28
14	3.33	3.88	3.70	4 31	3.75	4.37	4.17	4 86
13	3.81	4 44	4.73	4.93	4.23	4 93	4.70	5.48

NOTE.—The galvanizing of sheet iron adds about one-third of a pound to its weight per square foot.

AMERICAN AND BIRMINGHAM WIRE GAUGES.

No. Gauge	Thickness American Gauge	Thickness Birmingham Gauge	No. Gauge	Thickness American Gauge	Thickness Birmingham Gauge	No. Gauge	Thickness American Gauge	Thickness Birmingham Gauge
	Inch.	Inch.		Inch.	Inch.		Inch.	Inch.
0000	.46	.454	11	.0907	.12	25	.0179	.02
000	.4096	.425	12	.0808	.109	26	.0160	.018
00	.3648	.38	13	.0719	.095	27	.0142	.016
0	.3248	.34	14	.0641	.083	28	.0126	.014
1	.2893	.30	15	.057	.072	29	.0112	.013
2	.2576	.284	16	.0508	.065	30	.01	.012
3	.2294	.259	17	.0452	.058	31	.0089	.01
4	.2043	.238	18	.0403	.049	32	.0079	.009
5	.1819	.22	19	.0359	.042	33	.007	.008
6	.1620	.203	20	.0319	.035	34	.0063	.007
7	.1443	.18	21	.0284	.032	35	.0056	.005
8	.1285	.165	22	.0253	.028	36	.005	.004
9	.1144	.148	23	.0225	.025			
10	.1019	.134	24	.0201	.022			

RAILROAD SPIKES.

Length and Thickness in a Keg of 150 Pounds.

Length.	Thickness.	Number.	Length.	Thickness.	Number.
4⅞	$\frac{7}{16}$	527	5½	⅝	356
4½	$\frac{7}{16}$	400	5½	$\frac{9}{16}$	290
5		710	5⅝		219
5	$\frac{7}{16}$	489	6		311
5	½	390	6	$\frac{9}{16}$	263
5	$\frac{9}{16}$	296	6	⅝	197
5	⅝	258			

SPLICES AND BOLTS FOR ONE MILE OF TRACK.

Rails 30 feet long take 704 splices, 1408 bolts.
" 28 " " 754 " 1508 "
" 27 " " 782 " 1564 "
" 25 " " 844 " 1688 "
" 24 " " 880 " 1760 "

RAILROAD IRON.

To find the number of tons of rails for one mile of single track, divide the weight per yard by 7 and multiply by 11. Thus: for 56 lb. rail, $56 \div 7 = 8$, and $8 \times 11 = 88$ tons per mile.

WEIGHT OF ROLLED LEAD, COPPER, AND BRASS.—SHEET AND BAR.

Thickness or Diameter, in Inches.	LEAD			COPPER			BRASS			Thickness or Diameter, in Inches.
	Sheets per Square Foot.	Square Bars 1 foot long.	Round Bars 1 foot long.	Sheets per Square Foot.	Square Bars 1 foot long.	Round Bars 1 foot long.	Sheets per Square Foot.	Square Bars 1 foot long.	Round Bars 1 foot long.	
	Lbs.	Lbs.	Lbs.	Lbs.	Lbs.	Lbs.	Lbs.	Lbs.	Lbs.	
1-32	1.86	.005	.004	1.44	.004	.003	1.36	.004	.003	1-32
1-16	3.72	.019	.015	2.89	.015	.012	2.71	.014	.011	1-16
3-32	5.58	.044	.034	4.33	.034	.027	4.06	.032	.025	3-32
1-8	7.44	.078	.061	5.77	.060	.047	5.42	.056	.044	1-8
5-32	9.30	.121	.095	7.20	.094	.074	6.75	.088	.069	5-32
3-16	11.2	.174	.137	8.66	.135	.106	8.13	.127	.100	3-16
7-32	13.0	.237	.187	10.1	.184	.144	9.50	.173	.136	7-32
1-4	14.9	.310	.244	11.5	.240	.189	10.8	.226	.177	1-4
5-16	18.6	.485	.381	14.4	.376	.295	13.5	.353	.277	5-16
3-8	22.3	.698	.548	17.3	.541	.435	16.3	.508	.399	3-8
7-16	26.0	.950	.746	20.2	.736	.578	19.0	.691	.543	7-16
1-2	29.8	1.24	.974	23.1	.962	.755	21.7	.903	.739	1-2
9-16	33.5	1.57	1.23	26.0	1.22	.955	24.3	1.14	.900	9-16
5-8	37.2	1.94	1.52	28.9	1.50	1.18	27.1	1.41	1.11	5-8
11-16	40.9	2.34	1.84	31.7	1.82	1.43	29.8	1.70	1.34	11-16
3-4	44.6	2.79	2.19	34.6	2.16	1.70	32.5	2.03	1.60	3-4
13-16	48.3	3.27	2.57	37.5	2.55	1.99	35.2	2.38	1.87	13-16
7-8	52.1	3.80	2.98	40.4	2.94	2.31	37.9	2.76	2.17	7-8
15-16	56.0	4.37	3.42	43.3	3.35	2.65	40.6	3.18	2.49	15-16
1.	59.5	4.96	3.90	46.2	3.85	3.02	43.3	3.61	2.84	1.
1-8	66.9	6.27	4.92	52.0	4.87	3.82	48.7	4.57	3.60	1-8
1-4	74.4	7.75	6.09	57.7	6.01	4.72	54.2	5.64	4.43	1-4
3-8	81.8	9.37	7.37	63.5	7.28	5.72	59.6	6.82	5.37	3-8
1-2	89.3	11.2	8.77	69.3	8.65	6.80	65.0	8.12	6.38	1-2
5-8	96.7	13.1	10.3	75.1	10.2	7.98	70.4	9.53	7.49	5-8
3-4	104.	15.2	11.9	80.8	11.8	9.25	75.9	11.1	8.68	3-4
7-8	112.	17.5	13.7	86.6	13.5	10.6	81.3	12.7	9.97	7-8
2.	119.	19.8	15.6	92.3	15.4	12.1	86.7	14.4	11.3	2.

WIRE.

IRON, STEEL, COPPER, BRASS.

Weight of 100 Feet in Pounds. Birmingham Wire Gauge.

No. of Gauge.	PER LINEAL FOOT.			
	Iron.	Steel.	Copper.	Brass.
0000	54.62	55.13	62.39	58.93
000	47.86	48.32	54.67	51.64
00	38.27	38.63	43.71	41.28
0	30.63	30.92	34.99	33.05
1	23.85	24.07	27.24	25.73
2	21.37	21.57	24.41	23.06
3	17.78	17.94	20.3	19.18
4	15.01	15.15	17.15	16.19
5	12.82	12.95	14.65	13.84
6	10.92	11.02	12.47	11.78
7	8.586	8.667	9.807	9.263
8	7.214	7.283	8.241	7.783
9	5.805	5.859	6.63	6.262
10	4.758	4.803	5.435	5.133
11	3.816	3.852	4.359	4.117
12	3.148	3.178	3.596	3.397
13	2.392	2.414	2.732	2.58
14	1.826	1.843	2.085	1.969
15	1.374	1.387	1.569	1.482
16	1.119	1.13	1.279	1.208
17	.8915	.9	1.018	.9618
18	.6363	.6423	.7268	.6864
19	.4675	.472	.534	.5043
20	.3246	.3277	.3709	.3502
21	.2714	.274	.31	.2929
22	.2079	.2098	.2373	.2241
23	.1656	.1672	.1892	.1788
24	.1283	.1295	.1465	.1384
25	.106	.107	.1211	.1144
26	.0859	.0867	.0981	.0926
27	.0678	.0685	.0775	.0732
28	.0519	.0524	.0593	.056
29	.0448	.0452	.0511	.0483
30	.0382	.0385	.0436	.0412
31	.0265	.0267	.0303	.0286
32	.0215	.0217	.0245	.0231
33	.017	.0171	.0194	.0183
34	.013	.0131	.0148	.014
35	.0066	.0067	.0076	.0071
36	.0042	.0043	.0048	.0046

IRON RIVETS.

WEIGHT IN POUNDS PER 100.

Length Under Head, Inches.	DIAMETERS, INCHES.						
	$\frac{1}{4}$	$\frac{3}{8}$	$\frac{1}{2}$	$\frac{5}{8}$	$\frac{3}{4}$	$\frac{7}{8}$	1
	Lbs.	Lbs.	Lbs.	Lbs.	Lbs.	Lbs.	Lbs.
1	1.895	4.848	.966	16.79	26.49	39.3	55.2
	2.067	5.235	10.34	17.86	27.99	41.4	57.9
	2.238	5.616	11.04	18.96	29.61	43.5	60.7
	2.410	6.003	11.73	20.03	31.13	45.6	63.4
	2.582	6.402	12.43	21.04	32.74	47.8	66.2
	2.754	6.789	13.12	22.11	34.25	49.9	68.9
	2.926	7.179	13.81	23.21	35.86	52.0	71.7
	3.098	7.566	14.50	24.28	37.37	54.1	74.4
2	3.269	7.956	15.19	25.48	38.99	56.3	77.2
	3.441	8.343	15.88	26.56	40.40	58.4	79.9
	3.613	8.733	16.57	27.65	42.11	60.5	82.7
	3.785	9.120	17.26	28.73	43.67	62.6	85.4
	3.957	9.511	17.95	29.82	45.24	64.8	88.2
	4.129	9.898	18.64	30.90	46.80	66.9	90.9
	4.301	10.29	19.33	31.99	48.36	69.0	93.7
	4.473	10.67	20.02	33.08	49.92	71.1	96.4
3	4.644	11.06	20.71	34.18	51.49	73.3	99.2
	4.816	11.44	21.40	35.27	53.05	75.4	101.9
	4.988	11.84	22.09	36.35	54.61	77.5	104.7
	5.160	12.23	22.78	37.44	56.17	79.6	107.4
	5.332	12.62	23.48	38.52	57.74	81.8	110.2
	5.504	13.01	24.17	39.60	59.30	83.9	112.9
	5.676	13.39	24.86	40.69	60.86	86.0	116.7
	5.848	13.78	25.55	41.78	62.42	88.1	119.4
4	6.019	14.17	26.24	42.87	63.99	90.3	121.2
	6.191	14.56	26.93	43.94	65.55	92.4	123.9
	6.363	14.95	27.62	45.01	67.11	94.5	126.6
100 Heads.	.519	1.74	4.14	8.10	13.99	22.27	33.15

Length of rivet required to make one head = $1\frac{1}{2}$ diameters of round bar.

NAILS AND SPIKES.

Size, Length, and Number to the Pound.

CUMBERLAND NAIL AND IRON CO.

ORDINARY.

Size.	Length.	No. to Lb.
2^d	7/8	716
3 fine	1 1/16	588
3	1 1/8	448
4	1 3/8	336
5	1 3/4	216
6	2	166
7	2 1/4	118
8	2 1/2	94
10	2 3/4	72
12	3 1/4	50
20	3 3/4	32
30	4 1/4	20
40	4 3/4	17
50	5	14
60	5 1/2	10

LIGHT.

Size.	Length.	No. to Lb.
4^d	1 3/8	373
5	1 3/4	272
6	2	196

BRADS.

Size.	Length.	No. to Lb.
6^d	2	16?
8	2 1/2	96
10	2 3/4	74
12	3 1/8	50

CLINCH.

Length.	No. to Lb.
2	152
2 1/4	133
2 1/2	92
2 3/4	72
3	60
3 1/4	43

FENCE.

Length.	No. to Lb.
2	96
2 1/4	66
2 1/2	56
2 3/4	50
3	40

SPIKES.

Length.	No. to Lb.
3 1/2	19
4	15
4 1/2	13
5	10
5 1/2	9
6	7

BOAT.

Length.	No. to Lb.
1 1/2	206

FINISHING.

Size.	Length.	No. to Lb.
4^d	1 3/8	384
5	1 5/8	256
6	2	204
8	2 1/2	102
10	3	80
12	3 3/8	65
20	3 1/2	46

CORE.

Size.	Length.	No. to Lb.
6^d	2	143
8	2 1/4	68
10	2 3/4	60
12	3 1/4	42
20	3 5/8	25
30	4 1/4	18
40	4 3/4	14
W H	2 1/4	69
W H L	2 1/2	72

SLATE.

Size.	Length.	No. to Lb.
3^d	1 3/16	288
4	1 7/16	244
5	1 3/4	187
6	2	140

TACKS.

Size.	Length	Number to Pound.	Size.	Length	Number to Pound.	Size.	Length	Number to Pound.
1 oz.	1/8	16000	4 oz.	7/16	4000	14 oz.	13/16	1143
1 1/2	3/16	10066	6	9/16	2666	16	7/8	1000
2	1/4	8000	8	5/8	2000	18	15/16	888
2 1/2	5/16	6400	10	11/16	1600	20	1	800
3	3/8	5333	12	3/4	1333	22	1 1/16	727

UNITED STATES STANDARD SIZES

SQUARE AND HEXAGON NUTS.

Number of each size in 100 Lbs.

BLANK NUTS—NOT TAPPED.

Size of Bolt.	SIZE OF NUT.		SQUARE.		HEXAGON.	
	Width.	Thickness.	No. in 100 Lbs.	Weight each in Lbs.	No. in 100 Lbs.	Weight each in Lbs.
$\frac{1}{4}$	$\frac{1}{2}$	$\frac{1}{4}$	7400	.013	8880	.011
$\frac{5}{16}$	$\frac{19}{32}$	$\frac{5}{16}$	4000	.025	4800	.020
$\frac{3}{8}$	$\frac{11}{16}$	$\frac{3}{8}$	2730	.036	3276	.030
$\frac{7}{16}$	$\frac{25}{32}$	$\frac{7}{16}$	1700	.058	2040	.050
$\frac{1}{2}$	$\frac{7}{8}$	$\frac{1}{2}$	1160	.086	1392	.071
$\frac{9}{16}$	$\frac{31}{32}$	$\frac{9}{16}$	900	.111	1080	.092
$\frac{5}{8}$	$1\frac{1}{16}$	$\frac{5}{8}$	653	.153	784	.127
$\frac{3}{4}$	$1\frac{1}{4}$	$\frac{3}{4}$	386	.259	463	.215
$\frac{7}{8}$	$1\frac{7}{16}$	$\frac{7}{8}$	260	.384	312	.320
1	$1\frac{5}{8}$	1	170	.588	204	.490
$1\frac{1}{8}$	$1\frac{13}{16}$	$1\frac{1}{8}$	122	.819	146	.684
$1\frac{1}{4}$	2	$1\frac{1}{4}$	90	1.111	108	.925
$1\frac{3}{8}$	$2\frac{3}{16}$	$1\frac{3}{8}$	69	1.44	83	1.20
$1\frac{1}{2}$	$2\frac{3}{8}$	$1\frac{1}{2}$	54	1.85	65	1.53
$1\frac{5}{8}$	$2\frac{9}{16}$	$1\frac{5}{8}$	43	2.32	52	1.92
$1\frac{3}{4}$	$2\frac{3}{4}$	$1\frac{3}{4}$	35	2.85	42	2.38
$1\frac{7}{8}$	$2\frac{15}{16}$	$1\frac{7}{8}$	29	3.44	35	2.85
2	$3\frac{1}{8}$	2	24	4.16	30	3.33
$2\frac{1}{8}$	$3\frac{5}{16}$	$2\frac{1}{8}$	20	5.00	26	3.84
$2\frac{1}{4}$	$3\frac{1}{2}$	$2\frac{1}{4}$	17	5.88	22	4.54
$2\frac{3}{8}$	$3\frac{11}{16}$	$2\frac{3}{8}$	14	7.14	19	5.26
$2\frac{1}{2}$	$3\frac{7}{8}$	$2\frac{1}{2}$	12	8.33	16	6.25
$2\frac{3}{4}$	$4\frac{1}{4}$	$2\frac{3}{4}$	10	10.00	13	7.69
3	$4\frac{5}{8}$	3	8	12.50	10	10.00

BOLTS.

WITH SQUARE HEADS AND NUTS.

Weight of 100 of the Enumerated Sizes.

Lengths.	¼ in.	⅜ in.	½ in.	⅝ in.	¾ in.	⅞ in.	1 in.	1⅛ in.
Inch.								
1½	4.16	10.62	23.87	39.31				
1¾	4.22	11.72	25.06	41.38				
2	4.75	12.38	26.44	45.69	73.62			
2¼	5.34	12.90	28.62	49.50	76.			
2½	5.97	14.69	29.50	51.25	79.75			
2¾	6.50	16.47	31.16	53.	83.			
3	....	17.87	32.44	56.	85.38	127.25		
3½	....	18.94	39.75	63.12	93.44	140.56		
4	....	20.59	42.50	74.87	108.12	148.37	228.	296.
4½	....	21.69	44.87	79.62	113.12	158.76	239.	310.
5	....	23.62	48.81	83.	122.	167.25	250.	324.
5½	....	25.81	51.38	87.88	128.62	174.88	261.	338.
6	....	26.87	53.31	92.38	131.75	204.25	272.	352.
6½	....	....	56.87	96.88	139.56	214.69	283.	366.
7	....	....	59.12	99.87	145.50	228.44	294.	370.
7½	....	....	61.87	105.75	150.88	235.31	305.	384.
8	....	....	64.44	109.50	157.12	239.88	316.	398.
9	....	....	70.50	118.12	169.62	258.12	338.	426.
10	....	....	77.	128.13	184.	276.18	360.	454.
11	....	....	82.88	136.19	195.13	295.69	382.	482.
12	....	....	86.37	144.87	209.75	311.94	404	510.
13	....	....	92.	155.50	219.37	335.81	426.	538.
14	....	....	97.75	163.58	237.50	351.88	448.	566.
15	....	....	103.25	170.75	249.06	391.75	470.	594.

STANDARD SIZES OF WASHERS.

Number in 100 Pounds.

Diameter.	Size of Hole.	Thickness Wire Gauge.	Size of Bolt.	Number in 100 Lbs.
Inch.	*Inch.*	*No.*	*Inch.*	
⅝	5/16	16	¼	29300
¾	⅜	16	5/16	18000
1	7/16	14	⅜	7600
1¼	9/16	11	½	3300
1⅜	⅝	11	9/16	2180
1½	11/16	11	⅝	2350
1¾	13/16	11	¾	1680
2	15/16	10	⅞	1140
2½	1 1/16	8	1	580
2¾	1¼	8	1⅛	470
3	1⅜	7	1¼	360
3	1½	6	1⅜	360

WROUGHT-IRON WELDED TUBES, FOR STEAM, GAS, OR WATER.

1¼ inch and below, Butt Welded; proved to 800 lbs. per square inch, Hydraulic Pressure.
1½ inch and above, Lap Welded; proved to 500 lbs. per square inch, Hydraulic Pressure.

TABLE OF STANDARD SIZES. MORRIS, TASKER & CO.

Inside Diameter.	Actual Inside Diameter.	Actual Outside Diameter.	Thickness.	Internal Circumference.	External Circumference.	Length of Pipe per □ Foot, Inside Surface.	Length of Pipe per □ Foot, Outside Surface.	Internal Area.	External Area.	Length of Pipe containing 1 Cubic Foot.	Weight per Foot of Length.	No. of Threads per Inch of Screw.
Inches.	Inches.	Inches.	Inches.	Inches.	Inches.	Feet.	Feet.	Inches.	Inches.	Feet.	Lbs.	
⅛	.270	.405	.068	.848	1.272	14.15	9.44	.0572	.129	2500.	.243	27
¼	.364	.54	.088	1.144	1.696	10.50	7.075	.1041	.229	1385.	.422	18
⅜	.494	.675	.091	1.552	2.121	7.67	5.657	.1916	.358	751.5	.561	18
½	.623	.84	.109	1.957	2.652	6.13	4.502	.3048	.554	472.4	.845	14
¾	.824	1.05	.113	2.589	3.299	4.635	3.637	.5333	.866	270.	1.126	14
1	1.048	1.315	.134	3.292	4.134	3.679	2.903	.8627	1.357	166.9	1.670	11½
1¼	1.380	1.66	.140	4.335	5.215	2.768	2.301	1.496	2.164	96.25	2.258	11½
1½	1.611	1.9	.145	5.061	5.969	2.371	2.01	2.038	2.835	70.65	2.694	11½
2	2.067	2.375	.154	6.494	7.461	1.848	1.611	3.355	4.430	42.36	3.667	11½
2½	2.468	2.875	.204	7.754	9.032	1.547	1.328	4.783	6.491	30.11	5.773	8
3	3.067	3.5	.217	9.636	11.996	1.245	1.091	7.388	9.621	19.49	7.547	8
3½	3.548	4.	.226	11.146	12.566	1.077	.955	9.887	12.566	14.56	9.055	8
4	4.026	4.5	.237	12.648	14.137	.949	.849	12.730	15.904	11.31	10.728	8
4½	4.508	5.	.247	14.153	15.708	.848	.765	15.939	19.635	9.03	12.492	8
5	5.045	5.563	.259	15.849	17.475	.757	.629	19.990	24.299	7.20	14.564	8
6	6.065	6.625	.280	19.054	20.813	.63	.577	28.889	34.471	4.98	18.767	8
7	7.023	7.625	.301	22.063	23.954	.544	.505	38.737	45.663	3.72	23.410	8
8	7.982	8.625	.322	25.076	27.096	.478	.444	50.039	58.426	2.88	28.348	8
9	9.001	9.688	.344	28.277	30.433	.425	.394	63.633	73.715	2.26	34.077	8
10	10.019	10.75	.366	31.475	33.772	.381	.355	78.838	90.762	1.80	40.641	8

LAP WELDED
AMERICAN CHARCOAL IRON BOILER TUBES.
Tables of Standard Sizes.

External Diameter	Internal Diameter	Thickness	External Circumference	Internal Circumference	Length Pipe per Ft., Inside Surface	Length Pipe per Ft., Outside Surface	Internal Area	External Area	Weight per Foot
In.	In.	In.	In.	In.	Ft.	Ft.	In.	In.	Lbs.
1	0.856	0.072	3.142	2.689	4.460	3.819	0.575	0.785	0.708
1¼	1.106	0.072	3.927	3.474	3.455	3.056	0.960	1.227	0.9
1½	1.334	0.083	4.712	4.191	2.863	2.547	1.396	1.767	1.250
1¾	1.560	0.095	5.498	4.901	2.448	2.183	1.911	2.405	1.665
2	1.804	0.098	6.283	5.667	2.118	1.909	2.556	3.142	1.981
2¼	2.054	0.098	7.069	6.484	1.850	1.698	3.314	3.976	2.238
2½	2.283	0.109	7.854	7.172	1.673	1.528	4.094	4.909	2.755
2¾	2.533	0.109	8.639	7.957	1.508	1.390	5.039	5.940	3.045
3	2.783	0.109	9.425	8.743	1.373	1.273	6.083	7.069	3.333
3¼	3.012	0.119	10.210	9.462	1.268	1.175	7.125	8.296	3.958
3½	3.262	0.119	10.995	10.248	1.171	1.091	8.357	9.621	4.272
3¾	3.512	0.119	11.781	11.033	1.088	1.018	9.687	11.045	4.590
4	3.741	0.130	12.566	11.753	1.023	0.955	10.992	12.566	5.320
4½	4.241	0.130	14.137	13.323	0.901	0.849	14.126	15.904	6.010
5	4.72	0.140	15.708	14.818	0.809	0.764	17.497	19.635	7.226
6	5.699	0.151	18.849	17.904	0.670	0.637	25.509	28.274	9.346
7	6.657	0.172	21.991	20.914	0.574	0.545	34.803	38.484	12.435
8	7.636	0.182	25.132	23.989	0.500	0.478	45.795	50.265	15.109
9	8.615	0.193	28.274	27.055	0.444	0.424	58.291	63.617	18.002
10	9.573	0.214	31.416	30.074	0.399	0.382	71.975	78.540	22.19

WROUGHT-IRON WELDED TUBES.
Extra Strong.

Nominal Diameter	Actual Outside Diameter	Thickness, Extra Strong	Thickness, Double Extra Strong	Actual Inside Diameter, Extra Strong	Actual Inside Diameter, Double Extra Strong
⅛	.405	.100	......	.205	
¼	.54	.123	......	.294	
⅜	.675	.127	......	.421	
½	.84	.149	.298	.542	.244
¾	1.05	.157	.314	.736	.422
1	1.315	.182	.364	.951	.587
1¼	1.66	.194	.388	1.272	.884
1½	1.9	.203	.406	1.494	1.088
2	2.375	.221	.442	1.933	1.491
2½	2.875	.280	.560	2.315	1.755
3	3.5	.304	.608	2.892	2.284
3½	4.	.321	.642	3.358	2.716
4	4.5	.341	.682	3.818	3.136

WINDOW GLASS.

Number of Lights per Box of 50 Feet.

Inches.	No.	Inches.	No.	Inches.	No.	Inches.	No.
6 × 8	150	12 × 18	33	16 × 44	10	26 × 32	9
7 × 9	115	20	30	18 × 20	20	34	8
8 × 10	90	22	27	22	18	36	8
11	82	24	25	24	17	40	7
12	75	26	23	26	15	42	7
13	70	28	21	28	14	44	6
14	64	30	20	30	13	48	6
15	60	32	18	32	13	50	6
16	55	34	17	34	12	54	5
9 × 11	72	13 × 14	40	36	11	58	5
12	67	16	35	38	11	28 × 30	9
13	62	18	31	40	10	32	8
14	57	20	28	44	9	34	8
15	53	22	25	20 × 22	16	36	7
16	50	24	23	24	15	38	7
17	47	26	21	26	14	40	6
18	44	28	19	28	13	44	6
20	40	30	18	30	12	46	6
10 × 12	60	14 × 16	32	32	11	50	5
13	55	18	29	34	11	52	5
14	52	20	26	36	10	56	4
15	48	22	23	38	9	30 × 36	7
16	45	24	22	40	9	40	6
17	42	26	20	44	8	42	6
18	40	28	18	46	8	44	5
20	36	30	17	48	8	46	5
22	33	32	16	50	7	48	5
24	30	34	15	60	6	50	5
26	28	36	14	22 × 24	14	54	4
28	26	40	13	26	13	56	4
30	24	44	11	28	12	60	4
32	22	15 × 18	27	30	11	32 × 42	5
34	21	20	24	32	10	44	5
11 × 13	50	22	22	34	10	46	5
14	47	24	20	36	9	48	5
15	44	26	18	38	9	50	4
16	41	28	17	40	8	54	4
17	39	30	16	44	8	56	4
18	36	32	15	46	7	60	4
20	33	16 × 18	25	50	7	34 × 40	5
22	30	20	23	24 × 28	11	44	5
24	27	22	20	30	10	46	5
26	25	24	19	32	9	50	4
28	23	26	17	36	8	52	4
30	21	28	16	40	8	56	4
32	20	30	15	44	7	36 × 44	5
34	19	32	14	46	7	50	4
12 × 14	43	34	13	48	6	56	4
15	40	36	12	50	6	60	3
16	38	38	12	54	5	64	3
17	35	40	11	56	5	40 × 60	3

SKYLIGHT AND FLOOR GLASS.

Weight per Cubic Foot, 156 Pounds.

WEIGHT PER SQUARE FOOT.

Thickness .	$\frac{1}{8}$	$\frac{3}{16}$	$\frac{1}{4}$	$\frac{3}{8}$	$\frac{1}{2}$	$\frac{5}{8}$	$\frac{3}{4}$	1 inch.
Weight . .	1.62	2.43	3.25	4.88	6.50	8.13	9.75	13 lbs.

FLAGGING.

Weight per Cubic Foot, 168 Pounds.

WEIGHT PER SQUARE FOOT.

Thickness .	1	2	3	4	5	6	7	8 inch.
Weight . .	14	28	42	56	70	84	98	112 lbs.

CAPACITY OF CISTERN.

In Gallons, for each Foot in Depth.

Diameter, in Feet.	Gallons.	Diameter, in Feet.	Gallons.
2.	23.5	9.	475.87
2.5	36.7	9.5	553.67
3.	52.9	10.	587.5
3.5	71.96	11.	710.9
4.	94.02	12.	846.4
4.5	119.	13.	992.9
5.	146.8	14.	1151.5
5.5	177.7	15.	1321.9
6.	211.6	20.	2350.0
6.5	248.22	25.	3570.7
7.	287.84	30.	5287.7
7.5	330.48	35.	7189.
8.	376.	40.	9367.2
8.5	424.44	45.	11893.2

The American standard gallon contains 231 cubic inches, or 8$\frac{1}{3}$ pounds of pure water. A cubic foot contains 62.3 pounds of water, or 7.48 gallons. Pressure per square inch is equal to the depth or head in feet multiplied by .433. Each 27.72 inches of depth gives a pressure of one pound to the square inch.

ROOFING SLATE.

General Rule for the Computation of Slate.

From the length of the slate take three inches, or as many as the third covers the first; divide the remainder by 2, and multiply the quotient by the width of the slate, and the product will be the number of square inches in a single slate. Divide the number of square inches thus procured by 144, the number of square inches in a square foot, and the quotient will be the number of feet and inches required. A square of slate is what will cover 100 square feet, when laid upon the roof.

Weight per Cubic Foot, 174 Pounds.

WEIGHT PER SQUARE FOOT.

Thickness .	$\frac{1}{8}$	$\frac{3}{16}$	$\frac{1}{4}$	$\frac{3}{8}$	$\frac{1}{2}$	$\frac{5}{8}$	$\frac{3}{4}$	1 inch.
Weight . .	1.81	2.71	3.62	5.43	7.25	9.06	10.87	14.5 lbs.

TABLE OF SIZES AND NUMBER OF SLATE

In One Square.

Size, in Inches.	No. of Slate in Square.	Size, in Inches.	No. of Slate in Square.	Size, in Inches.	No. of Slate in Square.
6 × 12	533	8 × 16	277	12 × 20	141
7 12	457	9 16	246	14 20	121
8 12	400	10 16	221	11 22	137
9 12	355	12 16	184	12 22	126
10 12	320	9 18	213	14 22	108
12 12	266	10 18	192	12 24	114
7 14	374	11 18	174	14 24	98
8 14	327	12 18	160	16 24	86
9 14	291	14 18	137	14 26	89
10 14	261	10 20	169	16 26	78
12 14	218	11 20	154		

SPECIFIC GRAVITY
AND
WEIGHTS OF VARIOUS SUBSTANCES.

Name of Substance.	WEIGHTS.			Specific Gravity.
	Per Cubic Foot.	Per ☐ Foot, 1 In. Thick.	Per Cubic Inch.	
Water, Pure . . .	62.3	5.19	.036	1.000
Water, Sea . . .	64.3	5.36	.037	1.028
Wrought Iron . .	480	40.00	.277	7.70
Cast Iron	450	37.50	.260	7.20
Steel	490	40.84	.283	7.84
Lead	710	59.16	.410	11.36
Copper, Rolled . .	548	45.66	.317	8.80
Brass, Rolled . .	524	43.66	.302	8.40
Sand	98	8.23	.057	1.57
Clay	120	10.00	.069	1.92
Brickwork, Common	120	10.00	.069	1.92
" Close Joints	140	11.66	.081	2.24
Limestone . . .	168	18.00	.124	2.68
Glass	156	13.00	.090	2.49
Pine, White . . .	30	2.50	.017	.48
Pine, Yellow . . .	35	2.91	.019	.56
Hemlock	25	2.08	.015	.40
Maple	49	4.08	.028	.78
Oak, White . . .	50	4.16	.030	.80
Walnut	41	3.41	.023	.65

PROPERTIES OF CIRCLES.

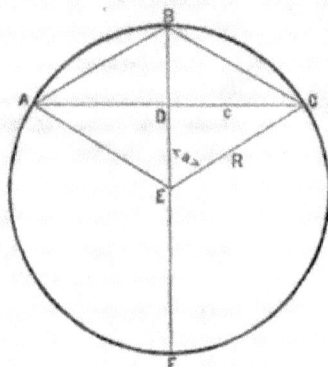

$$BD = h = R (1 - \cos. a)$$

$$\text{Sin.} \, a = \frac{\frac{1}{2} c}{R}$$

(1.) Given, chord A D C and vers. sine or rise B D, to find radius,

$$\frac{ADC}{2} = AD \text{ or } DC \therefore \frac{AD^2 + BD^2}{2\, BD} = BE$$

$$R = \frac{c^2 + 4\, h^2}{8\, h}$$

(2.) Given, chord A D C and radius B E, to find rise B D,

$$BE - \sqrt{BE^2 - AD^2} = BD$$

$$h = R - \sqrt{R^2 - \frac{c^2}{4}}$$

(3.) Given, the radius and rise, to find the chord A D C,

$$AD = \sqrt{BE^2 - (BE - BD)^2}$$

$$\text{Chord } ADC = 2\,AD = 2\sqrt{BE^2 - (BE - BD)^2}$$

$$c = 2\sqrt{2\, h\, R - h^2}$$

(4.) Given, the chord of an arc and the chord of half the arc, to find the length of the arc,

$$\frac{8\ A\ B - A\ D\ C}{3} = \text{arc A B C (very nearly).}$$

(5.) To find the number of degrees in the arc of a circle, when the diameter, or radius, and the length of the arc are given,

$$\frac{\text{Arc A B C}}{\pi \times \text{diameter}} \times 360° = \text{degrees in arc A B C}$$

(6.) Length of an arc of one degree $= R \times .0174533$
Length of an arc of one minute $= R \times .0002909$
Length of an arc of one second $= R \times .0000048$

Example.—Let radius $= 100$ feet, and the angle of the arc be 90°. What is the length of the arc?

$$100 \times .0174533 \times 90° = 157.08 \text{ feet.}$$

MENSURATION OF SURFACES.

Area of circle $=$ Diameter$^2 \times .7854$
Area of ellipse $=$ Transv. axis $\times$ conjug. axis $\times .7854$
Area of sector of circle $=$ Arc $\times \frac{1}{2}$ radius
Area of parabola $=$ Base $\times \frac{2}{3}$ height
Surface of sphere $=$ Diameter$^2 \times 3.1416$

MENSURATION OF SOLIDS.

Cylinder $=$ Area of one end $\times$ length
Sphere $=$ Diameter$^2 \times .5236$
Cone, or pyramid $=$ Area of base $\times \frac{1}{3}$ height
Any prismoid $=$ Sum of areas of the two parallel surfaces $+ 4$ times the area of a midway section $\times$ length, and the total product divided by 6.

PROPERTIES OF TRIANGLES.

In right-angled triangles

hypoth.^{2} = base^2 + perpend.^{2}
base^2 = (hyp. + perp.) $\times$ (hyp.—perp.)
perp.^{2} = (hyp. + base) $\times$ (hyp.—base)

VALUE OF ANY SIDE **A**.

$$A = \frac{B \sin. a}{\sin. b} \qquad\qquad A = \frac{C \sin. a}{\sin. c}$$

$$A = \sqrt{B^2 + C^2 - 2\, B\, C \cos. a}$$

$$A = \frac{B}{\cos. c + \sin. c \cot. a}$$

$$A = \frac{C}{\cos. b + \sin. b \cot. a}$$

$$A = B \cos. c + B \sin. c \cot. b$$

VALUE OF ANY ANGLE.

$$\sin. b = \frac{B \sin. a}{A} \qquad\qquad \sin. b = \frac{B \sin. a}{C}$$

$$\cos. b = \frac{A^2 + C^2 - B^2}{2\, A\, C}$$

$$\sin. b = \sin. (c + a).$$
$$\sin. b = \sin. c \cos. a + \cos. c \sin. a.$$

TRIGONOMETRICAL EXPRESSIONS.

The diagram shows the different trigonometrical expressions in terms of the angle A.

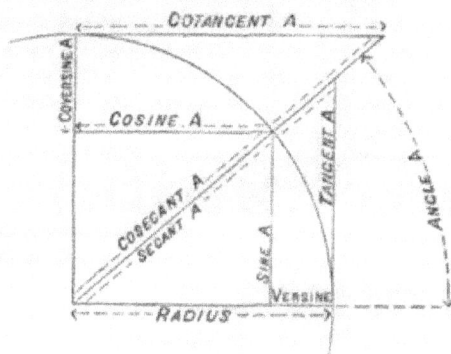

Complement of an angle = its difference from 90°.
Supplement . . . . = its difference from 180°.

TRIGONOMETRICAL EQUIVALENTS.

$\sqrt{(1-Sin^2)}$	= Cosin.	$\sqrt{(1-Cosin^2)}$	= Sine.
Sin ÷ Tan	= Cosin.	Cosin ÷ Cotan	= Sine.
Sin × Cotan	= Cosin.	1 ÷ Cotan	= Tangent.
Sine ÷ Cos	= Tangent.	1 ÷ Sin	= Cosecant.
Cos ÷ Sine	= Cotang.	1 ÷ Cosin	= Secant.
$Sin^2 + Cos^2$	= Rad^2.	1 ÷ Cosecant	= Sine.
$Rad^2 + Tan^2$	= $Secant^2$.	1 ÷ Secant	= Cosin.
1 ÷ Tan	= Cotang.	Rad — Cosin	= Versin.
		Rad — Sin	= Coversin.

USE OF TABLE OF NATURAL SINES, Etc.

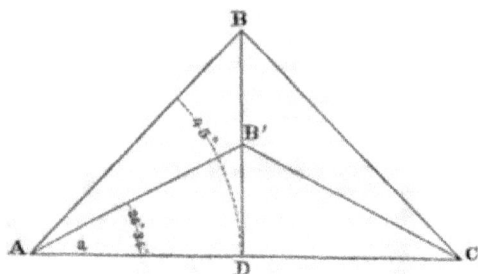

Example 1. To find the angle *a*, when A D and B′ D are given, from table of natural sines and tangents, p. 153.

A D being radius, B′ D = tan a.   Let $\begin{cases} \text{A D} = 20. \\ \text{B′D} = 10. \end{cases}$

Then $\dfrac{\text{B′ D}}{\text{A D}} = \dfrac{10}{20} = .50000.$

Referring to table we find for

    26°, the natural tangent to be .48773
    27°, the natural tangent to be .50952

       Difference 02179

The angle, therefore, is **more than** 26 and less than **27** degrees. If greater accuracy is required, take the difference between **natural** tangent of **26°** and **27°** as above, viz., .02179, and divide by **60, which** will give .00036 for one **minute.** Now subtract **from** .50000 the natural tangent for **26°, viz.,** .48773, leaving .01227, and divide the difference by .00036; the quotient will be 34 minutes. The angle, therefore, is 26° 34′.

Example 2. If A D = 20, and B D = **20,** what will be the angle subtended by B D?

$$\frac{\text{B D}}{\text{A D}} = \frac{20}{20} = 1.0000.$$

The natural tangent of 45° is 1.

NATURAL SINES, ETC.

Deg.	Sine.	Cover.	Cosecant	Tangent	Cotang.	Secant.	Versine.	Cosine.	Deg.
0	.00	1.00000	Infinite.	.0	Infinite.	1.00000	.0	1.00000	90
1	.01745	.98254	57.2986	.01745	57.2899	1.00015	.0001	.99984	89
2	.03489	.96510	28.6537	.03492	28.6362	1.00060	.0006	.99939	88
3	.05233	.94766	19.1073	.05240	19.0811	1.00137	.0013	.99862	87
4	.06975	.93024	14.3355	.06992	14.3006	1.00244	.0024	.99756	86
5	.08715	.91284	11.4737	.08748	11.4300	1.00381	.0038	.99619	85
6	.10452	.89547	9.5667	.10510	9.5143	1.00550	.0054	.99452	84
7	.12186	.87813	8.2055	.12278	8.1443	1.00750	.0074	.99254	83
8	.13917	.86082	7.1852	.14054	7.1153	1.00982	.0097	.99026	82
9	.15643	.84356	6.3924	.15838	6.3137	1.01246	.0123	.98768	81
10	.17364	.82635	5.7587	.17632	5.6712	1.01542	.0151	.98480	80
11	.19080	.80919	5.2408	.19438	5.1445	1.01871	.0183	.98162	79
12	.20791	.79208	4.8097	.21255	4.7046	1.02234	.0218	.97814	78
13	.22495	.77504	4.4454	.23086	4.3314	1.02630	.0256	.97437	77
14	.24192	.75807	4.1335	.24932	4.0107	1.03061	.0297	.97029	76
15	.25881	.74118	3.8637	.26794	3.7320	1.03527	.0340	.96592	75
16	.27563	.72436	3.6279	.28674	3.4874	1.04029	.0387	.96126	74
17	.29237	.70762	3.4203	.30573	3.2708	1.04569	.0436	.95630	73
18	.30901	.69098	3.2360	.32491	3.0776	1.05146	.0489	.95105	72
19	.32556	.67443	3.0715	.34432	2.9042	1.05762	.0544	.94551	71
20	.34202	.65797	2.9238	.36397	2.7474	1.06417	.0603	.93969	70
21	.35836	.64163	2.7904	.38386	2.6050	1.07114	.0664	.93358	69
22	.37460	.62539	2.6694	.40402	2.4750	1.07853	.0728	.92718	68
23	.39073	.60926	2.5593	.42447	2.3558	1.08636	.0794	.92050	67
24	.40673	.59326	2.4585	.44522	2.2460	1.09463	.0864	.91354	66
25	.42261	.57738	2.3662	.46630	2.1445	1.10337	.0936	.90630	65
26	.43837	.56162	2.2811	.48773	2.0503	1.11260	.1012	.89879	64
27	.45399	.54600	2.2026	.50952	1.9626	1.12232	.1089	.89100	63
28	.46947	.53052	2.1300	.53170	1.8807	1.13257	.1170	.88294	62
29	.48480	.51519	2.0626	.55430	1.8040	1.14335	.1253	.87461	61
30	.50000	.50000	2.0000	.57735	1.7320	1.15470	.1339	.86602	60
31	.51503	.48496	1.9416	.60086	1.6642	1.16663	.1428	.85716	59
32	.52991	.47008	1.8870	.62486	1.6003	1.17917	.1519	.84804	58
33	.54463	.45536	1.8360	.64940	1.5398	1.19236	.1613	.83867	57
34	.55919	.44080	1.7882	.67450	1.4825	1.20621	.1709	.82903	56
35	.57357	.42642	1.7434	.70020	1.4281	1.22077	.1808	.81915	55
36	.58778	.41221	1.7013	.72654	1.3763	1.23606	.1909	.80901	54
37	.60181	.39818	1.6616	.75355	1.3270	1.25213	.2013	.79863	53
38	.61566	.38433	1.6242	.78128	1.2799	1.26901	.2119	.78801	52
39	.62932	.37067	1.5890	.80978	1.2348	1.28675	.2228	.77714	51
40	.64278	.35721	1.5557	.83909	1.1917	1.30540	.2339	.76604	50
41	.65605	.34394	1.5242	.86928	1.1503	1.32501	.2452	.75470	49
42	.66913	.33086	1.4944	.90040	1.1106	1.34563	.2568	.74314	48
43	.68199	.31800	1.4662	.93251	1.0723	1.36732	.2685	.73135	47
44	.69465	.30534	1.4395	.96568	1.0355	1.39016	.2806	.71933	46
45	.70710	.29289	1.4142	1.00000	1.0000	1.41421	.2928	.70710	45

| | Cosine. | Versine. | Secant. | Cotang. | Tangent | Cosecant. | Cover. | Sine. | |

CIRCUMFERENCES OF CIRCLES.
Advancing by Eighths.

CIRCUMFERENCES.

Diam.	.0	⅛	¼	⅜	½	⅝	¾	⅞
0	.0	.3927	.7854	1.178	1.570	1.963	2.356	2.748
1	3.141	3.534	3.927	4.319	4.712	5.105	5.497	5.890
2	6.283	6.675	7.063	7.461	7.854	8.246	8.639	9.032
3	9.424	9.817	10.21	10.60	10.99	11.38	11.78	12.17
4	12.56	12.93	13.35	13.74	14.13	14.52	14.92	15.31
5	15.70	16.10	16.49	16.88	17.27	17.67	18.06	18.45
6	18.84	19.24	19.63	20.02	20.42	20.81	21.20	21.59
7	21.99	22.38	22.77	23.16	23.56	23.95	24.34	24.74
8	25.13	25.52	25.91	26.31	26.70	27.09	27.48	27.88
9	28.27	28.66	29.05	29.45	29.84	30.23	30.63	31.02
10	31.41	31.80	32.20	32.59	32.98	33.37	33.77	34.16
11	34.55	34.95	35.34	35.73	36.12	36.52	36.91	37.30
12	37.69	38.09	38.48	38.87	39.27	39.66	40.05	40.44
13	40.84	41.23	41.62	42.01	42.41	42.80	43.19	43.58
14	43.98	44.37	44.76	45.16	45.55	45.94	46.33	46.73
15	47.12	47.51	47.90	48.30	48.69	49.08	49.48	49.87
16	50.26	50.65	51.05	51.44	51.83	52.22	52.62	53.01
17	53.40	53.79	54.19	54.58	54.97	55.37	55.76	56.15
18	56.54	56.94	57.33	57.72	58.11	58.51	58.90	59.29
19	59.69	60.08	60.47	60.86	61.26	61.65	62.04	62.43
20	62.83	63.22	63.61	64.01	64.40	64.79	65.18	65.58
21	65.97	66.36	66.75	67.15	67.54	67.93	68.32	68.72
22	69.11	69.50	69.90	70.29	70.68	71.07	71.47	71.86
23	72.25	72.64	73.04	73.43	73.82	74.22	74.61	75.00
24	75.39	75.79	76.18	76.57	76.96	77.36	77.75	78.14
25	78.54	78.93	79.32	79.71	80.10	80.50	80.89	81.28
26	81.68	82.07	82.46	82.85	83.25	83.64	84.03	84.43
27	84.82	85.21	85.60	86.00	86.39	86.78	87.17	87.57
28	87.96	88.35	88.75	89.14	89.53	89.92	90.32	90.71
29	91.10	91.49	91.89	92.28	92.67	93.06	93.46	93.85
30	94.24	94.64	95.03	95.42	95.81	96.21	96.60	96.99
31	97.39	97.78	98.17	98.57	98.96	99.35	99.75	100.14
32	100.53	100.92	101.32	101.71	102.10	102.49	102.89	103.29
33	103.67	104.07	104.46	104.85	105.24	105.64	106.03	106.42
34	106.81	107.21	107.60	107.99	108.39	108.78	109.17	109.56
35	109.96	110.35	110.74	111.13	111.53	111.92	112.31	112.71
36	113.10	113.49	113.88	114.28	114.67	115.06	115.45	115.85
37	116.24	116.63	117.02	117.42	117.81	118.20	118.60	118.99
38	119.38	119.77	120.17	120.56	120.95	121.34	121.74	122.13
39	122.52	122.92	123.31	123.70	124.09	124.49	124.88	125.27
40	125.66	126.06	126.45	126.84	127.24	127.63	128.02	128.41
41	128.81	129.20	127.59	129.98	130.38	130.77	131.16	131.55
42	131.95	132.34	132.73	133.13	133.52	133.91	134.30	134.70
43	135.09	135.48	135.87	136.27	136.66	137.05	137.45	137.84
44	138.23	138.62	139.02	139.41	139.80	140.19	140.59	140.98
45	141.37	141.76	142.16	142.55	142.94	143.34	143.73	144.12

AREAS OF CIRCLES.
Advancing by Eighths.

AREAS.

Diam.	.0	⅛	¼	⅜	½	⅝	¾	⅞
0	.0	.0122	.0490	.1104	.1963	.3068	.4417	.6013
1	.7854	.9940	1.227	1.484	1.767	2.073	2.405	2.761
2	3.1416	3.546	3.976	4.430	4.908	5.411	5.939	6.491
3	7.068	7.669	8.295	8.946	9.621	10.32	11.04	11.79
4	12.56	13.36	14.18	15.03	15.90	16.80	17.72	18.66
5	19.63	20.62	21.64	22.63	23.75	24.85	25.96	27.10
6	28.27	29.46	30.67	31.91	33.18	34.47	35.78	37.12
7	38.48	39.87	41.28	42.71	44.17	45.66	47.17	48.70
8	50.26	51.84	53.45	55.08	56.74	58.42	60.13	61.86
9	63.61	65.39	67.20	69.02	70.88	72.75	74.66	76.58
10	78.54	80.51	82.51	84.54	86.59	88.66	90.76	92.88
11	95.03	97.20	99.40	101.6	103.8	106.1	108.4	110.7
12	113.0	115.4	117.8	120.2	122.7	125.1	127.6	130.1
13	132.7	135.2	137.8	140.5	143.1	145.8	148.4	151.2
14	153.9	156.6	159.4	162.2	165.1	167.9	170.8	173.7
15	176.7	179.6	182.6	185.6	188.6	191.7	194.8	197.9
16	201.0	204.2	207.3	210.5	213.8	217.0	220.3	223.6
17	226.9	230.3	233.7	237.1	240.5	243.9	247.4	250.9
18	254.4	258.0	261.5	265.1	268.8	272.4	276.1	279.8
19	283.5	287.2	291.0	294.8	298.6	302.4	306.3	310.2
20	314.1	318.1	322.0	326.0	330.0	334.1	338.1	342.2
21	346.3	350.4	354.6	358.8	363.0	367.2	371.5	375.8
22	380.1	384.4	388.8	393.2	397.6	402.0	406.4	410.9
23	415.4	420.0	424.5	429.1	433.7	438.3	443.0	447.6
24	452.3	457.1	461.8	466.6	471.4	476.2	481.1	485.9
25	490.8	495.7	500.7	505.7	510.7	515.7	520.7	525.8
26	530.9	536.0	541.1	546.3	551.5	556.7	562.0	567.2
27	572.5	577.8	583.2	588.5	593.9	599.3	604.8	610.2
28	615.7	621.2	626.7	632.3	637.9	643.5	649.1	654.8
29	660.5	666.2	671.9	677.7	683.4	689.2	695.1	700.9
30	706.8	712.7	718.6	724.6	730.6	736.6	742.6	748.6
31	754.8	760.9	767.0	773.1	779.3	785.5	791.7	798.0
32	804.3	810.6	816.9	823.2	829.6	836.0	842.4	848.8
33	855.3	861.8	868.3	874.9	881.4	888.0	894.6	901.3
34	907.9	914.7	921.3	928.1	934.8	941.6	948.4	955.3
35	962.1	969.0	975.9	982.8	989.8	996.8	1003.8	1010.8
36	1017.9	1025.0	1032.1	1039.2	1046.3	1053.5	1060.7	1068.0
37	1075.2	1082.5	1089.8	1097.1	1104.5	1111.8	1119.2	1126.7
38	1134.1	1141.6	1149.1	1156.6	1164.2	1171.7	1179.3	1186.9
39	1194.6	1202.3	1210.0	1217.7	1225.4	1233.2	1241.0	1248.8
40	1256.6	1264.5	1272.4	1280.3	1288.2	1296.2	1304.2	1312.2
41	1320.3	1328.3	1336.4	1344.5	1352.7	1360.8	1369.0	1377.2
42	1385.4	1393.7	1402.0	1410.3	1418.6	1427.0	1435.4	1443.8
43	1452.2	1460.7	1469.1	1477.6	1486.2	1494.7	1503.3	1511.9
44	1520.5	1529.2	1537.9	1546.6	1555.3	1564.0	1572.8	1581.6
45	1590.4	1599.3	1608.2	1617.0	1626.0	1634.9	1643.9	1652.9

SURVEYING MEASURE.

(LINEAL.)

Inches.		Feet.		Yards.		Chains.		Mile.
1.	=	.0833	=	.0278	=	.00126	=	.0000158
12.		1.		.333		.01515		.000189
36.		3.		1.		.04545		.000568
792.		66.		22.		1.		.0125
63360.		5280.		1760.		80.		1.

One knot or geographical mile = 6086.07 feet = 1855.11 metres = 1.1526 statute mile.

One admiralty knot = 1.1515 statute miles = 6080 feet.

LONG MEASURE.

Inches.		Feet.		Yards.		Poles.		Furl.		Mile.
1.	=	.083	=	.02778	=	.005	=	.000126	=	.0000158
12.		1.		.333		.0606		.00151		.0001894
36.		3.		1.		.182		.00454		.000568
198.		16½.		5½.		1.		.025		.003125
7920.		660.		220.		40.		1.		.125
63360.		5280.		1760.		320.		8.		1.

A palm = 3 inches.   A hand = 4 inches.
A span = 9 inches.   A cable's length = 120 fathoms.

FRENCH LONG MEASURE.

	Inches.	Feet.	Yards.	Miles.
Millimetre.....	.03937	.0033		
Centimetre....	.39368	.0328		
Decimetre.....	3.9368	.3280	.10936	
Metre..........	39.368	3.2807	1.09357	
Decametre....	393.68	32.807	10.9357	
Hectometre...	....	328.07	109.357	.062134
Kilometre.....	....	3280.7	1093.57	.621346
Myriametre...	....	32807.	10935.7	6.213466

SQUARE MEASURE.

Inches.	Feet.	Yards.	Perches.	Acre.
1. =	.00694 =	.000772 =	.0000255 =	.000000159
144.	1.	.111	.00367	.000023
1296.	9.	1.	.0331	.0002066
39204.	272¼.	30¼.	1.	00625
6272640.	43560.	4840.	160.	1.

100 square feet   = 1 square.

10 square chains = 1 acre.

1 chain wide   = 8 acres per **mile.**

1 hectare   = 2.471143 acres.

1 square mile $\begin{cases} = 27{,}878{,}400 \text{ square feet.} \\ = 3{,}097{,}600 \text{ square yards.} \\ = 640 \text{ acres.} \end{cases}$

Acres $\times$ .0015625   = square miles.

Square yard $\times$ **.00000323** = square miles.

Acres $\times$ 4840 = square yards.

Square yards $\times$ .0002066   = **acres.**

A section of land is 1 mile square, and contains 640 acres.

A square acre is 208.71 ft. at each side; or, 220 $\times$ 198 ft.

A square ½ acre is 147.58 ft. at each side; or, 110 $\times$ 198 ft.

A square ¼ acre is 104.355 ft. at each side; or, 55 $\times$ 198 ft.

A circular acre is 235.504 ft. in diameter.

A circular ½ acre is 166.527 **ft.** in diameter.

A circular ¼ acre is **117.752 ft.** in diameter.

FRENCH SQUARE MEASURE.

Square.	Square Inches.	Square Feet.	Square Yards.
Millimetre.....	.00154	.0000107	.000001
Centimetre....	.15498	.0010763	.000119
Decimetre.....	15.498	.1076305	.011958
Metre **or Cen.**	1549.8	10.76305	1.19589
Decametre....	154988.	1076.305	119.589
Hectare........	....	107630.58	11958 95
Kilometre.....	.38607▢mls	10763058.	1195895.
Myriametre...	38.607 ''	....	....

CUBIC MEASURE.

Inches.	Feet.	Yard.	Cubic Metres.
1.	= .0005788	= .000002144	= .000016386
1728.	1.	.03704	.028315
46656.	27.	1.	.764513

A CUBIC FOOT IS EQUAL TO

1728 cubic inches,	29.92208 U. S. liquid quarts,
.037037 cubic yard,	25.71405 U. S. dry quarts,
.803564 U. S. struck bushel	59.84416 U. S. liquid pints,
of 2150.42 cub. in.	51.42809 U. S. dry pints.
3.21426 U. S. pecks.	239.37662 U. S. gills.
7.48052 U. S. liquid gallons	.26667 flour barrel of 3
of 231 cubic in.	struck bushels,
6.42851 U. S. dry gallons of	.23748 U. S. liquid barrel
268.8025 cubic in.	of 31½ gallons.

A cubic inch of water at 62° Fahr. weighs 252.458 grains.
A cubic foot of water at 62° Fahr. weighs 1002.7   ounces.
A cubic yard of water at 62° Fahr. weighs 1692.   pounds.

FRENCH CUBIC OR SOLID MEASURE.

		Pint.	Quart.	Bush.	Cubic Inch.	Cu. Ft.
Centilitre......	Dry ...	.0181			} .61016	
	Liquid	.0211				
Decilitre.......	Dry ...	.1816	.0908		} 6.1016	
	Liquid	.2113	.1056			
Litre............	Dry ...	1.816	.908		} 61.016	.0353
	Liquid	2.113	1.056			
Decalitre......	Dry ...		9.08	.2837	} 610.16	.3531
	Liquid	21.13	10.56			
Hectolitre....	Dry ...		90.8	2 837	} 6101.6	3.531
	Liquid	211.3	105.6			
Kilolitre or	Dry ...			28.37	} 61016.	35.31
Cubic Metre...	Liquid		1056.5			
Myriolitre.....	Dry ...			283.7	}	353.1
	Liquid		10565.			

AVOIRDUPOIS WEIGHT.

The standard avoirdupois pound is the weight of 27.7015 cubic inches of distilled water, weighed in the air, at 39.83 degrees Fahr., barometer at thirty inches.

Ounces.	Pounds.	Quarters.	Cwts.	Ton.
1. =	.0625 =	.00223 =	.000558 =	.000028
16.	1.	.0357	.00893	.000447
448.	28.	1.	.25	.0125
1792.	112.	4.	1.	.05
35840.	2240.	80.	20.	1.

A drachm = 27.343 grains.
A stone = 14 pounds.
A quintal = 100 kilogrammes.

7000 grains = 1 avoir. pound = 1.21528 troy pounds.
5760 grains = 1 troy pound = .82285 avoir. pound.

Kilos p. sq. centim. × 14.22 = Pounds p. sq. inch.
Pounds p. sq. inch × .0703 = Kilos p. sq. centim.

FRENCH WEIGHTS.

EQUIVALENT TO AVOIRDUPOIS.

	Grains.	Ounces.	Pounds.
Milligramme	.015433		
Centigramme......	.154331	.000352	.000022
Decigramme	1.54331	.003527	.000220
Gramme............	15.4331	.035275	.002204
Decagramme......	154.331	.352758	.022047
Hectogramme.....	1543.31	3.52758	.220473
Kilogramme........	15433.1	35.2758	2.20473
Myriogramme	....	352.758	22.0473
Quintal.............	....	3527.58	220.473
Millier or Tonne..	....	35275.8	2204.73